t, cameras, maps, mattresses, books, pic
lue, hammers, book covers, ata
elds, locks, silk-screens, arm ould
s, inflatables, rope, linotype machines, bo
vents, spanners, tyres, gaffer tape, ring-b
opied pictures, carnival masks, black cos
tones, photographs, catalogues, sleepin
lets, sandpaper, water pistols, helmets, g
rveys, banners, paper, pneumatic drills, p
tents, sandbags, carriages, murals, card
s, leaflets, wrought-iron, photographic pc
pes, police cars, movable type, bedstead.
aples, paintings, cooking grills, buses, mc
oi, pamphlets, sandbags, casements, lad
avans, tarpaulins, acetates, monuments,
ences, squeegees, cooking burners, sinks
riters, carbon paper, trees, bodies, beams
gaffer tape, rulers, small placards, windo
wspaper boxes, silhouettes, public billbo

RE/MAX
FOREVER
BU BİR
SİVİL DİREN
HİÇ BİR ŞE
ZARAR VE
EĞİTİM SİS
İYİLEŞTİRİL
TALEP EDİYOR
Topçu kışlası

DISOBEDIENT OBJECTS

Disobedient Objects

EDITED BY
CATHERINE FLOOD
& GAVIN GRINDON

V&A Publishing

First published by V&A Publishing, 2014
Victoria and Albert Museum
South Kensington
London SW7 2RL

ISBN: 978 1 85177 797 6
Library of Congress Catalog Control Number 2014932321
10 9 8 7 6 5 4 3 2 1
2017 2016 2015 2014

A catalogue record for this book is available from the British Library.

Designed by Marwan Kaabour at Barnbrook
'How-to' illustrations by Marwan Kaabour
Copy-edited by Denny Hemming
Index by Hilary Bird
New V&A photography by Richard Davis, George Eksts
and Jaron James, V&A Photographic Studio

Francesco Raparelli's essay is a version of a talk delivered in London in 2011.
David Graeber's essay is a highly condensed version of an essay that appears in *Possibilities: Essays on Hierarchy, Rebellion and Desire* (Oakland, 2007).

Printed in China

V&A Publishing

Supporting the world's leading
museum of art and design,
the Victoria and Albert
Museum, London

This book is typeset using Doctrine, designed by Jonathan Abbott, Jonathan Barnbrook and Julián Moncada. Doctrine is inspired by the livery and branding of Air Koryo, North Korea's flag carrier. The peculiar part political philosophy/part corporate branding conceptual mix inspired the development of the typeface. Choosing Doctrine for this book is intended to be in the same spirit of subverting everyday objects to perform disobedient functions.

Inside front cover: People gather around a barricade made from a bus at an entrance to Taksim Square, Istanbul, June 2013.

Frontispiece: A Palestinian youth aims a slingshot at an Israeli army jeep on the outskirts of Hebron during the Second Intifada, October 2000.

p.129: Allercombe camp, built to stop the construction of the A30 road in the west of England. The camp blocked construction for two years before it was evicted in 1997.

Inside back cover: People occupy a crane during protests against construction of the Newbury bypass, England, April 1996.

CONTENTS

'To disobey in order to take action is the byword of all creative spirits. The history of human progress amounts to a series of Promethean acts. But autonomy is also attained in the daily workings of individual lives by means of many small Promethean disobediences, at once clever, well thought out, and patiently pursued, so subtle at times as to avoid punishment entirely ... I would say that there is good reason to study the dynamics of disobedience, the spark behind all knowledge.'

—Gaston Bachelard, 'Prometheus', *Fragments of a Poetics of Fire*, 1961

Just inside the grand entrance to the V&A, on a balcony looking down on the exhibition this book accompanies, there is a sculpture of Hercules. It is one of many images of him in the museum – he appears not only in Greek and Roman sculptures and pottery, but also resurfaces in eighteenth-century oil paintings, fine ceramics and silver presentation vases which, according to the V&A's catalogue record, would 'have been displayed on the dining table during a very grand dinner'. In their history of the revolutionary Atlantic of the seventeenth and eighteenth centuries, Peter Linebaugh and Marcus Rediker observe that for the classically educated architects of the Atlantic economy, Hercules represented power and order. They saw in his mythical labours their own epic imperial ambitions and aggressive economic enclosure of the world. Accordingly, they placed his image on coins, buildings and the finely crafted objects of their domestic lives. Hercules' second labour was to destroy the Hydra of Lena, in whose image leaders of state and industry saw an antithetical figure of resistance and 'disorder'. It was an unruly monster, part whirlwind, part woman, part snake. When Hercules sliced off one of its heads, two more sprang up in its place. Eventually he killed it and, dipping arrows into the slain beast's gall, harnessed its power for himself and his future triumphs:

> From the beginning of English colonial expansion in the early seventeenth century through the metropolitan industrialization of the early nineteenth, rulers referred to the Hercules-hydra myth to describe the difficulty of imposing order on increasingly global systems of labor. They variously designated dispossessed commoners, transported felons, indentured servants, religious radicals, pirates, urban laborers, soldiers, sailors, and African slaves as the numerous, ever-changing heads of the monster. But the heads, though originally brought into productive combination by their Herculean rulers, soon developed among themselves new forms of cooperation against those rulers, from mutinies and strikes to riots and insurrections and revolution.[1]

Opposite Top Cup and saucer from a tea set with the emblem of the Women's Social and Political Union, H.M. Williamson & Sons, Bridge Pottery • Bone china with printed transfers • Longton, Britain, *c.*1910 • V&A: C.37C–1972 (cup), C.37D–1972 (saucer).

Opposite Bottom *Cacerola* lid, stainless steel, used in Buenos Aires, December, 2001. When the Argentinean government froze the bank accounts of 18 million citizens, thousands banged pots in the streets and shouted the slogan 'All of them must go!', forcing out four presidents in three weeks. It has been called the first national revolt against contemporary deregulated capitalism.

For Linebaugh and Rediker, the Hydra suggests, in silhouette, the lost history of the multi-ethnic classes essential to the making of the modern world. Historians like them have tried to look at history from below, instead of from the perspective of 'great men' and the agency of state and capital. History is inevitably a matter of selective inclusion. This is equally true of the objects of art and design history, whose collection is most often shaped by a market of wealthy collectors, even as some critical artists, curators and historians have attempted to intervene within the field. In that inevitable taking of sides, our project turns to objects that open up histories of making from below. These objects disclose hidden moments in which, even if only in brief flashes, we find the possibility that things might be otherwise: that, in fact, the world may also be made from below, by collective, organized disobedience against the world as it is.

But history from below can be difficult to perceive. Its protagonists are barely documented, and we can only tell so much by turning things like silver vases inside out in order to reveal them in negative relief. The art, design and material culture of these other classes went mostly uncollected, unpreserved, excluded from their place in the making of history. We know less of how these rebels represented themselves and their oppressors through objects and images than we do about the representations of Hercules collected by the V&A, after its founding in 1852 at the height of this struggle for enclosure. Culture, understood (in one narrow sense) as the objects and images we should know about and value – our history of art and design – is also often told from above. This exhibition is one for the Hydra.[2]

FIRST THEY IGNORE YOU. THEN THEY RIDICULE YOU. AND THEN THEY ATTACK YOU AND WANT TO BURN YOU. AND THEN THEY BUILD MONUMENTS TO YOU.

— Nicholas Klein, 'Address,' *Proceedings of the Third Biennial Convention of the Amalgamated Clothing Workers of America*, Baltimore, Maryland, 18 May 1919

Seeing through the Hydra's eyes is often a matter of historical perspective. Social movements,[3] whether focused on feminism, anti-capitalism, global justice or other issues, are at the centre of the struggles that have won many of the rights and liberties we now enjoy. They establish new ways of seeing the world and relating to each other that are often later taken for granted. Social movements are one of the primary engines producing our culture and politics, and this is no less true when it comes to art and design.

WHAT ARE DISOBEDIENT OBJECTS?

Disobedient objects have a history as long as social struggle itself. Ordinary people have always used them to exert 'counterpower'.[4] Objects have played a key role in social change alongside performance, music and the visual arts. Here the focus is on the previously under-examined area of the art and design of object-making within social movements, a people's history of art and design from below. Yet the imagination and creativity of making within social movements has played a key role in achieving social change; upending the terms of public debates; and directly influencing more familiar commercial art and design. The role of material culture in social movements is a mostly untold story. There have been many exhibitions of political prints, and there have been exhibitions of movement histories, mostly in social history museums, which included objects but did not focus specifically on them and their making.[5]

Likewise, writing on movement cultures has focused on print, performance or music, but less often on object-making.

Social movements, though they may appear chaotic, are one of the principal sites where culture grows. The most common lazy stereotypes, easy to find in certain newspapers, of movements as insensible, unthinking or inevitably violent, draw on even older classist, racist and sexist Victorian tropes of the flighty, swinish multitude or childlike, colonial savage, which have their roots in a bourgeois fear of the urban poor and 'oriental' culture. Little better is the notion of movement cultures as mechanisms of blunt political demands (as in the crudely statist notion of 'propaganda'). Instead, we borrow McPhee and Greenwald's phrasing of 'social movement cultures', which consciously identifies them as a site of culture and value.[6]

Our research converged on the question of movement objects. One of us (an art historian specializing in activist-art) was interested in the contours of how some aspects of social movement cultures have been included in institutions and

Opposite Top Dockers Union Export Branch banner, George Tutill for the Dock, Wharf, Riverside and General Workers Union of Great Britain and Ireland • Jacquard-woven silk with oil-painted images • Britain, early 1890s • People's History Museum NMLH.1993.580 • This union banner reappropriates the symbolism of Hercules' battle with the Hydra for labour movements. Inverting the symbolism of the Trafalgar Vase, the serpent here represents capitalism's constrictive enclosure of a powerful worker's body. Similar Herculean images would form part of labour iconography for many years to come.

Opposite Middle The Trafalgar Vase, Scott Digby • Silver • London, 1805-6 • V&A: 803:1-2–1890.

Top Cheap Art Manifesto, Bread and Puppet Theater • Print on paper • Vermont, 1985 • V&A E.45–2014

histories of art and design, while others have been excluded. Movement cultures are the zero-point of political art, but tend to be alternately ignored or problematically recuperated by art and design institutions. Formally, music and performance emerging from social movements have received perhaps the most attention from writers, curators and film-makers; the material objects of movements have most often fallen beyond their remit. Institutions have an understanding of what constitutes good design based on criteria of aesthetic excellence rooted in self-perpetuating professional infrastructures and ideas of connoisseurship. The V&A, for example, has mostly collected commodity-objects of elite production and consumption – also primarily objects of private consumption. An exception is collections of prints and posters. The multiple, cheap and distributed nature of the poster means that even in its most finely designed form it has been integrated into everyday public life. From the late nineteenth century museums began collecting posters, precisely because their public context suggested an exciting modern medium. A form that has commonly been used by activists (especially from the late 1960s) was therefore already an established museum object-type. So it is as prints and posters that movement cultures have most easily slipped under the doors of museums. For one of us (a curator of prints) it was the presence of protest graphics in the V&A collections that prompted thinking about the absence within the Museum of other kinds of disobedient objects.

There are many ways art and design practices can be politically active. But we aren't primarily concerned with the institutional frames of the sometimes isolated gestures of either 'critical design' or even programmes of 'interventionist' participatory art.[7] Likewise 'activist-art' and more recently 'design activism' are established terms referring respectively to a nebulously broad range of artists' practices or to top-down socially responsible professional design.[8] We do not wish to denigrate such practices, and it is true that there are many kinds of 'activism', but at the same time the broad use of the term 'activism' has also functioned as an enclosure of cultural value, authenticity and impact on the part of professional artists, critics, designers, corporations and even NGOs. Rather, it seems imperative to begin with the actually existing but often unacknowledged grassroots cultures of activist social movements in order to properly contextualize the many overlapping current debates on art, design and social change.

Yet even taking this focused notion of activism risks erasing differences: the relative strengths and weaknesses of the politics on display, and variations in the power, privilege and access of different movements. 'Activist objects' might suggest a narrow typology of objects made by 'activists',

Top A *cacerolazo* protest, Buenos Aires, Argentina, June 2008. Made in support of farmers striking against increased taxation, road blockades and noisemaking protests like this led the increases to be withdrawn.

an identity that does not always appropriately describe the forms of subjectivity involved in non-Western social movements. We also wished to acknowledge the micro-politics of the everyday, where social change is made before or beyond the composition of a recognizably 'activist' subjectivity.[9] For these reasons, we have avoided this more obvious term. Likewise, these aren't 'left' or 'right' wing objects. That rigid geometric scheme, which originates from the seating arrangements of the 1789 French National Assembly, is insufficiently nuanced to capture the diversity of movement cultures. Rather, these objects appear in varying, complexly composed movements, in which liberation movements may also be nationalist; deploy traditional, even religious, values; or oppose ostensibly 'left' communist states.

At the same time, 'disobedient objects' doesn't attempt to define a discipline. The term is intended as an evocative proposition or an invitation rather than a typology or closed concept. We look instead at the range of object-based tactics and strategies that movements adopt to succeed. Its edges remain open to questions. What other forms of agency do these objects involve? Can we identify material points where disobedience begins, or turns into something else? Are some politics unable to produce objects? We begin in the last years of the 1970s. Firstly, for practical reasons: many objects from before this period no longer exist, having been lost or destroyed, and have only been haphazardly documented in texts, photographs or films. The introductory section of the exhibition nonetheless includes a few key historical disobedient objects for context. Secondly, while the few years before 1980 seem a rather arbitrary beginning, they offer a means to start not with the crises of 1989, but with the swell of a global cycle of struggles that preceded them.[10] The objects made within social movements from this period to the present are not only bound to neo-liberal economic policies enacted on a broad scale from 1978,[11] but also to parallel changes in the organization and technology of work, leisure, communication and cultural production.

The earliest objects in our chronology in this respect are Chilean *arpilleras* (see p.122). In 1970 several United States corporations identified the democratic government of Chile as a problem. It was limiting their production and circulation of commodity-objects, from Pepsi bottles to the copper ingots melted to make the computer technologies then beginning to be woven into our lives. They asked the CIA to overthrow the Allende government. The CIA worked covertly to destabilize the country politically and economically and gave support to Pinochet's coup and the genocidal military dictatorship that followed. Meanwhile, Pinochet's US-trained economists used Chile to experiment with the then-untested economic ideas of Milton Friedman.[12] *Arpilleras* were objects on the other side of this history. Smuggled illegally out of Chile, they use traditional folk arts to simply and honestly make public the regime's torture camps and mass 'disappearances', and tell stories of women's everyday lives and resistance. In planning the Chilean coup, President Nixon instructed the

CIA to 'make the economy scream'.[13] The *arpilleras*, in their act of making and their depiction of murders alongside sunrises over the Andes mountains, embody both a scream of negation and a thread of hope for another future.[14]

WE WANT BREAD, BUT ROSES TOO

Lawrence Textile Strikers, attributed, 1921

There is no protest aesthetic. Political movement is always a matter of being emotionally moved, but each movement has its own aesthetic composition.[15] Accordingly the objects emerging from these cultures aren't unified by style or type. They can be monuments, full of symbolic historical accumulation, or small, quotidian and domestic. As much as they are often playful and humorous, they can also be simultaneously traumatic, traversed by antagonism and conflict. Their makers commonly experience pressure from governments and private economic interests, in the form of police harassment, violence, spying, imprisonment, even assassination.

The question of the value of these objects, not least in terms of beauty and aesthetic fineness, is starkly posed when these objects are placed in a museum such as the V&A. Displayed beside the V&A's examples of extravagantly fine craft, disobedient objects might seem to fail in comparative judgements of aesthetic quality. But a failure to pass can be a form of disobedience in its own right, not least in questioning the narrow grounds of 'quality'. Fine making often belongs to privileged social conditions involving time, institutional training, normalization and patronage. It is bound to discipline and governance. As a result, fine objects are themselves mostly failures in the task of making change.

Disobedient objects explore what Halberstam calls the queer art of failure.[16] They may be simple in means, but they are rich in ends. Working (in the words of Critical Art Ensemble) by any medium necessary, often under conditions of duress and scarcity, they tend to foreground promiscuous resourcefulness, ingenuity and timely intervention. This is not to balance aesthetic quality against social significance, but to begin to rethink aesthetic value itself. As Duncombe and Lambert argue: 'Political art ... is engaged in the world. The world is messy. It has lots of moving parts. This material is impossible to fully control or master ... Whereas compromise for the traditional artist means diluting their vision, compromise for the political artist is the very essence of democratic engagement.'[17]

Sometimes, however, a fine craft finish is exactly what allows an object to disobey. Carrie Reichardt, who works with ceramics and mosaic, maintains that the assumed politeness of the

Top Tree-sit, Pureora Forest, New Zealand, 1978. Following these protests, the government abandoned widespread logging of native trees.

medium means you can get away with more. There is a power in the double take that occurs between form and content:

> The beauty of craft is that at first it can seduce its audience. People are drawn in by the sheer skill and time taken to create a piece. I believe this allows a dialogue to open up where the viewer can be challenged both emotionally and intellectually. There is an expectation that craft work is gentle, decorative and safe – but once an audience is engaged it is the ideal place to explore radical and controversial ideas.[18]

The Bread and Puppet Theater has since the 1960s been central to introducing puppetry to social movements in the United States. Through the pathos of its archetypal papier-mâché puppets and Cheap Art Manifesto it negates stereotypes about social-movement making as crude or naïve because the objects are produced quickly, under pressure and with limited resources. Rather, movement makers are skilful artists, craftspeople and technologists producing considered, practical responses to complex problems, which have proven both effective and aesthetically powerful.

I PONDERED ALL THESE THINGS, AND HOW MEN FIGHT AND LOSE THE BATTLE, AND THE THING THAT THEY FOUGHT FOR COMES ABOUT IN SPITE OF THEIR DEFEAT, AND WHEN IT COMES TURNS OUT NOT TO BE WHAT THEY MEANT, AND OTHER MEN HAVE TO FIGHT FOR WHAT THEY MEANT UNDER ANOTHER NAME.

William Morris, *A Dream of John Ball*, 1888

The strange, sometimes ambivalent or bitter victories of movements complicate any assessment of successful design in their objects. Some disobedient objects might seem like 'hope in the dark', in Rebecca Solnit's phrase,[19] isolated projects unlikely to achieve widespread change. But their acts of composing things otherwise, in defiance of all that is wrong around them, are beautiful failures that throw teleological definitions of success into question. Moreover, all successful movements are made up of very large numbers of people carrying out small, seemingly utopian experiments without seeing or even necessarily knowing of each other; having no idea of the sometimes unlikely opportunities their acts might create; not necessarily realizing they are already sewing the fabric of historical change.

While the organizations that produce disobedient objects might have little cultural visibility to begin with, social movements are instituent – they aim to institute new ways of living, new laws and new social organizations. As William Morris observed, social movements often find themselves woven into unexpected new contexts that obscure their origins. Or as David Graeber puts it, 'What reformers have to understand is that they're never going to get anywhere without radicals and revolutionaries to betray.'[20] In Bolivia, the Katarista movements of the 1970s revived the Wiphala flag symbolizing Qullasuyu, their quadrant of the Inca empire, as part of their rural, indigenous and anti-colonial politics. The rainbow flag of forty-nine squares recalls pre-Columbian designs and became widespread in indigenous mobilizations in the 1990s. But between 2007 and 2009, when a new constitution refounded the country, the Wiphala's resonances altered as it became an official state flag, draped on government buildings and stitched to the uniforms of police and soldiers.[21]

If governments sometimes claim credit for movement victories and appropriate their established cultures, businesses more often do so with their cultural innovations. Today's proliferation of rentable public bicycles in cities began in Amsterdam with a collection of 1960s anarchist-artists called the Provos, who left white bicycles in public spaces for anyone to use and then leave for others. The police confiscated them, saying people might steal them (some Provos

responded by stealing police bikes, painting them white and leaving them out, too). Their white bike plan eventually led to government-supported bicycle programmes, since adopted by other city governments around the world. Similarly, the problematic labelling of the recent Arab Spring as the 'Twitter revolution' belies another genealogy: Twitter itself was inspired by an activist media project, the Institute for Applied Autonomy's TXTMob, launched (alongside the Ruckus Society's RNC Text Alert Service) to circumvent mass media and connect demonstrators during the 2004 Republican National Convention in New York (these initiatives were in turn inspired by early experiments with mobile phones and text messaging by European movements in the 1990s, especially Reclaim the Streets in Britain).

MAKING TROUBLE: SWARM DESIGN AND ECOLOGIES OF AGENCY

Disobedient objects are most commonly everyday objects appropriated and turned to a new purpose, from the wooden shoe of the saboteur (from *sabot*, French for wooden shoe) thrown into a factory machine to the shoe thrown at President Bush by an Iraqi journalist during a press conference with the words, 'This is a farewell kiss from the Iraqi people, you dog.'[22] Collective appropriation can be found in the noisemaking pots and pans first used in Chile's *cacerolazos* in the 1970s, in which the archetypal objects of domestic design sounded a counter public sphere, or the mass jingling of keys, which unlocked the air of public space during the 1989 Czech Velvet Revolution.[23] But disobedient objects are about making as much as breaking. Disobedience can involve DIY hacking and alteration, and also the design of whole new ways of disobeying. The re-use of easily accessible objects, like the shipping barrels composing nineteenth-century barricades (from *barrique*, French for barrel), implicate these objects in unfinished dialectics of social struggle and make them one means of the global circulation of struggles. For example, wooden pallets, the structural foundation of one unit load, were produced by the mid-twentieth-century standardization of international container shipping. They were brought about by efficiency drives rooted in de-skilling and breaking the power of unionized longshoremen's labour. But these mass-produced wooden frames, designed for disciplining labour and circulating commodities, became, around the world, a shared infrastructural basis for the first 1970s tree-sits in New Zealand (see p.12); furniture and barricade elements in 1970s Kabouter squats in the Netherlands, or those of Okupa in Spain; and more recently the base of 123 Occupy's designs to support the protest-unit of Occupy Wall Street tents (p.40).

Top Wiphala flag, La Paz, Bolivia, 2008.

Disobedient objects are not mere props. Or rather, as disability scholars have observed, democracy has always relied on prostheses. The system of voting, for example, has always been propped up by objects, from the Chartists' call for the democratizing impairment of secret ballots, where paper cards replace voices, to the push-button electronic voting machines introduced in India in the 1980s, which facilitated voting for illiterate citizens. Social movements, too, have their own props and they can fall down without them. (Even though, in British ecological movements, the key material infrastructures of protest events are referred to, self-depreciatingly, as 'activist tat'.) Though we have avoided the term, we might think of these as 'activist objects' in the sense that they are active, bound up with the agency of social change. The objects do not possess agency in themselves, but make change as part of ecologies composed also of other objects, music, performing bodies, technology, laws, organizations and affects. A weaker, less resourced power can triumph through asymmetrical innovation, and since the 1980s the strategic advantages of smallness and mobility have increased. So while disobedient objects are often appropriated, they also often appropriate their context of existing architecture or situations, unlocking them to reframe a situation or produce new relationships. As many have argued, the best response to a powerful enemy can be a more powerful story. Eclectic Electric Collective's inflatable cobblestones thrown at the police playfully destabilize relations between police and protesters (see p.73). The Book Bloc (discussed by Francesco Raparelli) implicates the police in a dance with demonstrators. The police's attempt to control the streets using violence is reframed as an attack on access to education. The holes wrought in the shields by the police's truncheons are part of their provenance, a certifying signature of their unwitting co-authorship.

While their social and geographical contexts vary widely, disobedient objects share common modes of production, lines of communication and influence. History from below entails multiplicity, and we focus on the interweaving of different historical moments. These objects don't move from producer to market in a circulation of commodities, as in Marx's scheme of **M**oney-**C**ommodity-**M**oney,[24] but are one means of a circulation of struggles (perhaps, Movement-Object-Movement). Making a new world is always an experiment, but it doesn't happen in an isolated laboratory. The objects involved are prototypes that exist in the wild, to be modified and reworked to meet the needs of different times and places. They have a distributed collective authorship, involving multiple reappropriations and reworkings as movements learn from each other and develop each other's tactics, or solve similar problems with parallel approaches.

TRIPODS

Tripods, objects that augment the body's ability to blockade, are an archetypal example of this swarm design. On 26 March 1974, loggers arrived in the village of Reni in Uttarakhand, northern India. Female villagers, after trying to reason with them, explaining that they relied on the trees for their livelihood, were threatened with guns. In response, they extended Ghandian methods to *chipko*: hugging the trees in a bodily blockade. Their successes in forest conservation became a strategic rallying point for the nascent ecological movement. In 1978 in New Zealand, as part of anti-logging protests that led to the foundation of Pureora Forest Park, activists extended such blockades by moving out of easy reach, building platforms using wooden pallets high up in the trees to blockade the felling with 'tree-sits', a tactic also adopted in Australia's Terrania Creek in 1979 (in what became national park land, including the picturesque Protesters Falls), and in the US in 1985 to prevent logging in Willamette National Forest, Oregon. As the tactic spread, tree surgery businesses or industrial rope access firms were sometimes hired in the United States and Britain to assist police and bailiffs in extracting protesters from trees. But protesters out-designed the authorities once again. In 1989, during huge anti-logging blockades in Coolangubra State Forest, Australia, activists raised a three-legged tripod about six metres high that blocked the single logging road into the forest: a tree-sit without a tree. The first tripod was a metal scaffold, pulled into place by a vehicle, but others there and at the parallel Chaelundi forest blockades used wooden logs (see p.65). One person sat atop the tripod, so that removing any of its legs would cause him or her to fall and be injured.[25] Some of these forests later became national parks. The North East Forest Alliance's 1991 *Intercontinental Deluxe Guide to Blockading* spread tripod (and lock-on) designs to the UK and US (some individual activists travelled between Australian, American and British actions, too). In the US, wooden tripods first appeared in 1992 blockades protecting the Cove Mallard wilderness. In Britain, the tripod was adopted by Reclaim the Streets, where urban activists with strong ties to earlier British tree-sits scavenged steel scaffolding poles to make tripods. In an urban context they constituted 'intelligent barricades' that closed a road to cars but left it open for pedestrians and bicycles. Beginning on Angel High Street, London, in 1994, these tripods made Reclaim the Streets parties possible. The design spread through the how-to guide *Road Raging*. Bipod and even unipod designs, alongside complex multi-tripod architectural arrangements using overlapping legs, sometimes in response to the development of specialized police removal units, proliferated in the United States, Asia-Pacific and Europe. Groups invested in lighter, more quickly erected aluminium (and even bamboo) poles over steel scaffolding. From the 2006 British Climate Camp protests, the tripod became a graphic icon of protest and was sometimes erected at camp entrances for purely symbolic reasons.

This ecology of agency also involves different contexts and power relations, traversing and transforming these objects. The role of the law is perhaps the clearest example. The state, in a paradox of sovereignty, attempts to define what are legal and acceptable forms of protest against it. Many modern forms of action, such as unions or strikes, were once illegal and required either secrecy or open lawbreaking. Recently in Britain new laws redefining 'public order', as well as cuts to legal aid and investment of public money in the surveillance and disruption of peaceful movements, have curtailed the right to protest.[26] Objects are intimately involved in this negotiation, back and forth, of what constitutes the space of 'legitimate' protest. David Graeber's essay details the reclassification – as 'potential weapons' – of sticks that support a dancing puppet (perhaps explaining a move to inflatables). In Britain, the 1994 Criminal Justice Act, Section 60, made wearing a mask at a protest (for example, in objection to police data-gathering teams) an offence. In 2012 United Arab Emirates police announced that people should not wear Guy Fawkes masks as 'objects deemed to instigate unrest are illegal', while their import into Bahrain was banned in 2013. The Molotov cocktail, which first appeared during the Spanish Civil War and later in Finnish resistance to Soviet invasion in the 1930s, entailed a semi-permanent change in the status of mass-produced glass bottles as unproblematic everyday objects. In Belfast during the early 1990s art students carrying milk bottles (which they used to wash their paint brushes) were often stopped as potential terrorists because – for the state – their artists' tools had become irrevocably associated with more insurgent appropriation. Here, too, we must include the many imaginary disobedient objects that have been conjured by the police, and fed to the media, which have at various points served as a pretext for curtailing protests. Despite their potent psychological associations, these objects never surfaced at protests and would have little practical reason for doing so – from condoms filled with urine at the Seattle 1999 WTO protests[27] to 'rioters armed with samurai swords and machetes' at the London 2001 May Day protests.[28]

Sometimes the media's imaginative framing of objects is embraced by, or definitive for, movements, from the fictional 'bra burning' in reports on the 1968 Miss America protests[29] to the coinage of the term 'Black Bloc' by the German press in the 1980s to describe the dress of some Autonomen.

Top *Pouwhenua* carried by Cyril Chapman (Ngapuhi tribe) in Hikoi (Maori protest march) for land rights, Lambton Quay, Wellington, 13 October 1975. The *Pouwhenua* is a land marker whose base will not touch the ground until all land illegally acquired from Maori is returned.

THE MASTER'S TOOLS WILL NEVER DISMANTLE THE MASTER'S HOUSE.

Audre Lorde, *Sister Outsider*, 1984

Context is everything. We should be wary of any uncritical affirmation of the power of making, 'creative' activism or transversal innovation in the context of the neo-liberal relations of the 'creative industries'. Rather, the contradiction remains open: to produce any value at all capital relies on the same capacity to be creative that is always also escaping and refusing. Even the most ultra-left or experimental politics are indebted to the creativity of social movement cultural traditions. This creativity can come from mobilizing folk-traditions and religious or spiritual values: for example, in British labour union banners' dense iconography; Indonesian group Taring Padi's protest puppets' adaptation of the traditions of *wayang* puppet theatre; Muneteru Ujino's neo-folk metal Mikoshi used in the 2003 demonstrations in Japan against the invasion of Iraq (see p.121); the carved Maori *pouwhenua* (pre-European land marker post) made for carrying at the head of the 1975 Maori land rights march and subsequent protests; or the avatar of the Broom-Wielding Goddess of Good Governance (Swachha Narayani) protecting street hawkers in Sewa Nagar market in Delhi, who, in her many arms, holds a video camera to film the police.

While some peace movements have taken up the Biblical phrase 'swords into ploughshares', many more pacifist and playful disobedient objects only function in specific social-democratic contexts, in which governments, even if in increasingly limited ways, recognize people as subjects with a right of resistance to speak and act politically. Without such acknowledgement – most often the case for movements in the global South or composed of people of colour and indigenous communities – struggles for rights and freedoms sometimes necessarily take different forms, from urban self-defence to rural or desert guerrilla warfare.

Their objects necessarily become improvised objects of physical force, often outmatched by but dialectically bound to the violence and oppression they resist: from 'fards', single-shot guns made by blacksmiths from scrap water pipes, used by neighbourhood protection groups in the poorest areas of the 2011 Egyptian revolution, to 'technicals', the improvised battle-vehicles engineered by anti-Gadaffi rebels during the 2011 Libyan revolution.

UNDISCIPLINED KNOWLEDGE

I JUMPED UP AND SAID 'ARM ME – I'LL KILL A WHITE DUDE RIGHT NOW!' THE WHOLE [BLACK PANTHER] MEETING GOT QUIET. THEY CALLED ME TO THE FRONT OF THE ROOM, AND THE BROTHER WHO WAS RUNNING THE MEETING LOOKED AT ME FOR A MINUTE, AND THEN REACHED INTO THE DESK DRAWER. MY HEART WAS POUNDING. I WAS LIKE, 'OH MY GOD, HE'S GOING TO GIVE ME A BIG-ASS GUN!' AND HE HANDED ME A STACK OF BOOKS ... I SAID, 'EXCUSE ME, SIR, I THOUGHT YOU WERE GOING TO ARM ME?' HE SAID, 'I JUST DID.'

Jamal Joseph, interview in *Time* magazine, 9 February 2012

Disobedient objects also lead us to think about how movements produce new forms of knowledge and strategy that help us see from below. While they may find footholds in various disciplines, they also draw from popular global and local traditions of making, outside professional art and design or academia.[30] Some of these are evoked by the many how-to publications which instruct their readers on the design of disobedience: the barricade diagrams of Auguste Blanqui's 1866 *Instructions for an Insurrection*; Bread and Puppet Theater's *68 Ways to Make Really Big Puppets*; Dave Foreman's *Ecodefense: A Field Guide to Monkeywrenching*; *The Squatter's Handbook*; *The Activist Tat Collective Recipe Book* for camps and convergences, or the recent collection *Beautiful Trouble*. These objects embody knowledge and skills. They are not formed from nothing. We might consider the section of Marx's *Grundrisse*, in which he argues that the fixed capital of factory machines materially embodies the 'general intellect' of workers – their aggregate skill and knowledge – in order to replace them.[31] This might prompt us to wonder what other anti-capitalist machines the general intellect might imagine and embody itself

in. We might think of the objects and performances of social movements as just such machines, embodying knowledge otherwise. There is certainly a mutiny of professional knowledge, including design, in these objects. But they are also moulded by the collective, informal, experiential knowledge of local laws around protest; how to negotiate with police; political meeting and street protest dynamics. Additionally, they spring from a base in leisure and domestic skills that become political tools, from camping to knitting and sewing. Behind the design of tripods stand other changes in leisure and education, for example, the growth of climbing as a sporting activity and the growth of indoor walls in the 1980s, often appearing first in university gyms. Such knowledges are one example of what Harney and Moten call 'the undercommons'. Its appearance in the museum echoes its role in the university:

Opposite *Delia Smith's Basic Blockading*, anonymous • Print on paper • Britain, n.d. • Private collection.

Top 'Spear' presented in a police press conference, one day after a demonstration supporting the printers' strike at Wapping, east London, 24 January 1987. It is actually, as the hanging loop indicates, a traditional supporting pole for a union banner. The Haldane Lawyers' Society report on policing of this demonstration records eyewitness accounts of groups of officers confiscating and ripping up union banners, removing the poles and retreating behind police lines.

Above This banner was created by *Taller Popular de Serigrafía* (*TPS*) in 2005 as part of the National Movement for a 6-Hour Working Day in Argentina. Underground rail workers' historic 6-hour day (limited due to the work's intensity, and previously removed only by the 1973–90 dictatorship) was lost when underground rail was privatized in the 1990s, but reinstated following strikes in 2004. In 2005, with high unemployment and crashing wages, a movement began for a 6-hour day for all. Its slogan was 'less work so we can all work'. The banner is seen here with train worker unionists speaking in 2005.

> It cannot be denied that the university is a place of refuge and it cannot be accepted that the university is a place of enlightenment. In the face of these conditions one can only sneak into the university and steal what one can. To abuse its hospitality, to spite its mission, to join its refugee colony, its gypsy encampment, to be in but not of – this is the path of the subversive intellectual in the modern university.[32]

In this sense, too, we take the Museum at its word to truly be a public institution. This project also enters a series of current tangential discussions at the V&A. A new team of contemporary architecture, product design and digital curators are formulating a collecting policy that addresses design, politics and public life – shifting the emphasis from understanding what a professional designer does to realizing the impact that design has on the way we live.[33] *Disobedient Objects* enters these conversations, challenging the Museum by confronting it with objects that demand to be treated differently. The project has been described to us as institutional critique and there is inevitably some truth in this. It prompts the question of whether the Museum can resist the urge to recuperate these objects. In the nineteenth century it was claimed that museums could prevent riots and sedition (as well as drunkenness) by mopping up working-class leisure time.[34] What happens when you place disobedient objects at the heart of a building that was conceived for such obedient purposes?

The position of this project, both 'within and against' an institution, emerges principally from careful attention to these objects and their own instituent power.[35] Our project isn't just about antagonism, although that is important. Rather, it entails a 'with and for'. As a project's spaces of autonomy develop, less time might be spent in antagonism than in co-research towards a collective project, composing the many 'yeses' behind the 'noes'. In this exhibition we returned, in one sense, to a quite traditional idea of the etymological roots of a curator as one who cares. 'Care' is here used not in the sense of bureaucratic administration or discipline, but as an ethics of solidarity, mutual aid, even love.[36] Caring for these objects entails being with and for them, listening to them and understanding how their making is bound to a making of history that is both neglected and incomplete.

UNFINISHED OBJECTS

THE POSTERS PRODUCED BY THE ATELIER POPULAIRE ARE WEAPONS IN THE SERVICE OF THE STRUGGLE AND ARE AN INSEPARABLE PART OF IT. THEIR RIGHTFUL PLACE IS IN THE CENTRES OF CONFLICT, THAT IS TO SAY, IN THE STREETS AND ON THE WALLS OF THE FACTORIES. TO USE THEM FOR DECORATIVE PURPOSES, TO DISPLAY THEM IN BOURGEOIS PLACES OF CULTURE OR TO CONSIDER THEM AS OBJECTS OF AESTHETIC INTEREST, IS TO IMPAIR BOTH THEIR FUNCTION AND THEIR EFFECT. THIS IS WHY THE ATELIER POPULAIRE HAS ALWAYS REFUSED TO PUT THEM ON SALE. EVEN TO KEEP THEM AS HISTORICAL EVIDENCE OF A CERTAIN STAGE IN THE STRUGGLE IS A BETRAYAL, FOR THE STRUGGLE ITSELF IS OF SUCH PRIMARY IMPORTANCE THAT THE POSITION OF AN 'OUTSIDE' OBSERVER IS A FICTION WHICH INEVITABLY PLAYS INTO THE HANDS OF THE RULING CLASS. THAT IS WHY THESE WORKS SHOULD NOT BE TAKEN AS THE FINAL OUTCOME OF AN EXPERIENCE, BUT AS AN INDUCEMENT FOR FINDING, THROUGH CONTACT WITH THE MASSES, NEW LEVELS OF ACTION, BOTH ON THE CULTURAL AND THE POLITICAL PLANE.

Atelier Populaire statement, 1968

A long, familiar shadow is still cast by the outdated modernist framework of the 'museum as mausoleum' – places where objects go to die, where they are preserved as reference points in an authoritative scheme of the universe. It has been argued that in their need to make sense of all the things they contain, museums deny their essential heterogeneity and follow an impulse to flatten and homogenize the objects they display.[37] But swarm-designed objects are necessarily rough, raw things, whose edges are open to further modification and appropriation. Only their contexts of use make them whole, and this makes these objects unfinished in another, more teleological, sense. Rather than being 'dead' like a butterfly enclosed in a case, disobedient objects on view in an exhibition are unfinished, like a political sticker never stuck, its hope and rage still held fast to its laminate backing. Their aura is that of an unfulfilled promise. But this incompleteness needn't be a melancholy sign of failure so much as one of possibility.

A suffragette tea set promoting votes for women is a comfortable object to contemplate because a consensus has formed about the struggle that produced it – what happened, who won and what that means. The jeopardy, trauma and grief encapsulated in many contemporary disobedient objects, however, is raw and ongoing in ways that may make them uncomfortable or disturbing. They embody uncomfortable truths about the present and destabilize the official line of politicians and media organizations. They are full of uncertainty – and the empowering and terrifying idea that our own actions (and inaction) could make a difference.

Before we located them, some of these objects were retired from the street to rest unknown in private lofts or social centre basements.[38] Now they find themselves returned to visible public history to speak to us. For other objects, the disputes and struggles they represent have not ended, and when this exhibition ends they will return to take their place within them. Whatever our emotional reaction or identification with these unfinished objects, we mostly encounter them only for a brief moment, and even then always mediated by other objects and

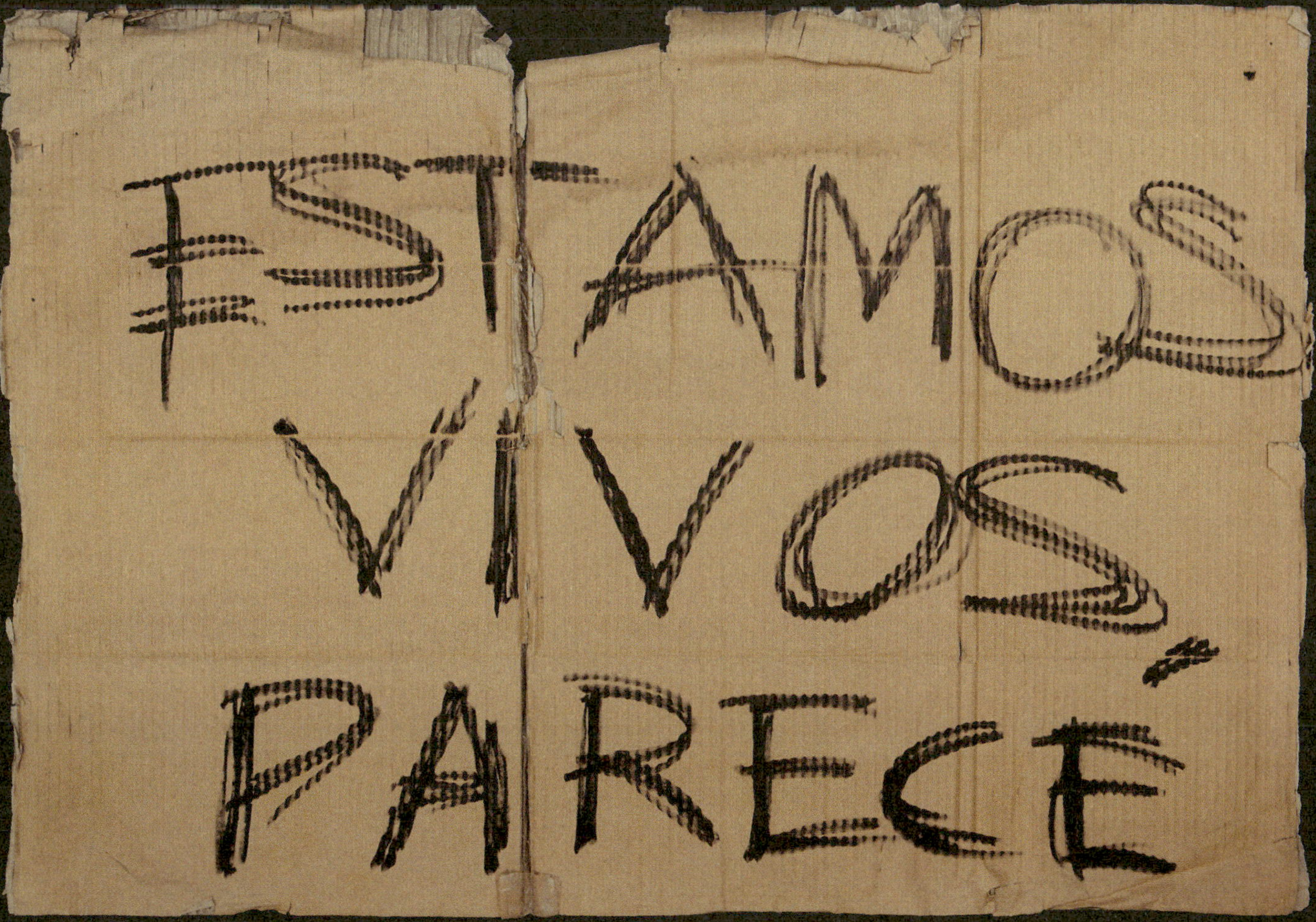

Top Placard • Black ink on cardboard • Madrid, 2011 • M15 Archive • Produced during the Acampada Sol M15 movement mobilizations, it reads 'we're alive, it seems'.

social relations: perhaps inches from, or touching, our bodies in a crowd; held by (or holding up) our friends or comrades; in news footage of people who could be us; in photographs of days growing distant; or suddenly reappearing in a courtroom. The exhibition of these objects is, in fact, one moment when you might actually spend time with them, right in front of you, able to slowly examine them. How does this moment of exhibition (where the objects are placed in historical, and relative, contexts) relate to these other moments, the object's use by activists, newspaper photographers, and so on? When objects such as these have appeared in museums, they have usually been presented as ephemera, displayed not for close attention in their own right so much as incidental objects that were present while important social change was happening. More rarely, they have appeared as fetishes, valorized as 'edgy' or 'vital' cultural capital and thus commodified in ways counter to the political goals they were made to achieve. These two conditions, ephemera and fetish, are the principal dangers we have tried to avoid.

The Atelier Populaire's resistance to institutionalization intersects with the anecdote quoted earlier from a Black Panther meeting, which suggests that reflection can be as important as action. But the terms of that reflection are crucial, and this problem of representation must be the primary concern when re-presenting social movement objects. The Atelier Populaire's critique, though totalizing, is well-founded. Social movements, in contesting our ways of seeing and acting, find themselves beset by a long and recent history of misrepresentation, in which they are ignored or maligned by mass media while simultaneously being appropriated for their vitality and authenticity. Museums are not immune to this process of caricature. Visiting the Political Art Documentation and Distribution archive at MoMA in New York, two independent researchers found a collection of undocumented American Indian movement posters, with a Post-It note inside their archive drawer that read, 'not cool enough to catalog'. Other groups, such as the Laboratory of Insurrectionary Imagination or the various Occupy movements, have found themselves invited – as content – to participate in museum programmes. The museum then often attempts to contain or stifle the same organizing vitality that originally attracted it when it becomes apparent that such organizing might trouble the museum's sponsorship or labour relations.[39]

Disobedient objects were not made with a museum in mind. Nor do they rely on the museum to legitimate them – but this does not mean that they have nothing to gain from appearing there. That an exhibition can provide space to consider, away from the rush of a political action or the hyperbole of mass media, was demonstrated at the ARTPLAY design space in Moscow during the demonstrations (or 'fair election' rallies) against Putin's election as president in 2012. Recognizing that a new style of public protest was emerging in Russia, exemplified by individualized and often witty handmade placards (see p.89), ARTPLAY invited protesters to lend their placards to the gallery for a short period during which they staged an exhibition, entitled *You don't even represent us / You can't even imagine us.* Afterwards many of the placards were collected by their makers and carried in further demonstrations. The exhibition was an affirmation that something significant was happening in terms of both politics and design. It marked a moment of birth rather than death for the objects.

Exhibitions are moments of collective meaning-making. Bringing these objects and histories together, and presenting them to an audience that never encounters them outside mass media, makes the museum a site for difficult questions and tests its claim to be a public space. But talking about movements outside the reach of those movements always involves discomfort. Rather than assuming a straightforward opposition between radical integrity and institutional separation, we attempted a more grounded approach to the re- and de-composition of these always-unfinished objects in relation to the making of movements. Movements begin already traversed by compromising power relations, and at certain points large institutions have also powerfully and honestly (as well as unknowingly) supported their development. Contemporary exhibitions have been a space of both possibilities and problems, paralleling the problems of radical history or philosophy texts that find themselves steeped in obfuscating language on inaccessibly expensive academic presses, or the contradictions of commercial distribution in which political documentary film can find itself. Invited to a dance with the institution that Holmes calls a game of liar's poker,[40] we set a wager on what the museum does to disobedient objects and what disobedient objects do to the museum. Just as troubling as the notion of museum-as-mausoleum is a newer metaphor that has emerged for the museum in a globalized free-market economy – the museum as supermarket, presenting the illusion of free choice. In the 'blockbuster' exhibition, the museum has perfected its own mode of cultural consumerism. Within a vast building teeming with all the possibilities of its permanent collections, visitors are presented with an exhibition event that is carefully explained (some would argue dumbed down), packaged and branded.[41] In asking how the museum might resist this kind of reduction, Pierre-Olivier Rollin has envisaged a different kind of exhibition 'where the visitor is invited to develop creative processes instead of consumerist habits'. Such an exhibition would be non-directive: 'there should be room for hesitations, backtracking and alternative routes ... it is of fundamental importance that visitors may not agree with the contents of the exhibition'. The responsibility of the museum, he suggests, is 'to organise an exhibition that is "controversial" at every level and that is permanently being "negotiated" by each individual visitor'.[42] It is within this kind of open-ended (many-headed) dialogic structure that we imagine disobedient objects might be able to come in to the museum and keep breathing.

Opposite This farmer in Hubei Province, China, resisted confiscation – by 100 riot-equipped government bailiffs – of his land for building projects in 2010 by using homemade rockets made of PVC pipe and fireworks in an elevated 'cannon tower'. He kept his land and sold it for record compensation.

Right This banner, pictured at a mass meeting in St Petersburg, 10 December 2011, employs a slogan by Pavel Arsenyev of the Laboratory of Poetic Actionism: '*Vy nas dazshe ne predstavliaete*' ('You don't even represent us' / 'You cannot even imagine us'). '*Predstavliaete*' means both 'imagine' and 'represent', evoking at once the failure of representative democracy, the misrepresentations of the media and the creative power of movements. It became iconic and was used continuously in demonstrations against Putin's government and election fraud from 4 December 2011 into 2012.

In thinking about the spatial organization of the *Disobedient Objects* exhibition, the starting point has been the idea that these artefacts are defined not by their form, but by their political efficacy. Their disobedience only becomes apparent when they are considered in context. To group these objects according to their formal qualities would iron out their particular contexts and reasons for being. There is a danger in conflating very different situations and experiences because the objects they produce look similar. In all instances we have attempted to present objects that are particular rather than typical.

The first four sections of the exhibition address the design logic of disobedient objects in relation to four strategies for social change: Direct Action, Speaking Out, Making Worlds and Solidarity. These four themes act as anchors around which the objects gravitate, encouraging visitors to make connections between objects and strategies of protest without these necessarily being mapped out. Introduced through quotations expressing these approaches, these strategies overlap and sometimes contradict each other. For example, there is a potential dichotomy between direct action and speaking out – what's the point of talking if you don't take action? What's the point of action if no one understands it?

Direct Action presents objects used in the empowering act of making change now, rather than asking political representatives to do so for you through mediated channels. From strikes to blockades, sit-ins and occupations, this sometimes means breaking the law, sometimes not. Direct action often involves blocking or slowing power, using objects such as the lock-on, which has its own history like that of the tripod. Speaking Out looks at how social movements get their message across when they are often misrepresented by mainstream media or are subject to censorship. Here we have traditional guerrilla communication (hand-painted placards, defaced currency) reinvigorated by social media, alongside tactical media experiments with new technologies. Making Worlds addresses the physical infrastructures of protest, which often embody 'prefigurative politics': anticipating new ways of living and relating to each other. These include the large-scale temporary structures of protest camps as well as small objects that provide support to protesting bodies, from bust cards to makeshift tear-gas masks. In Solidarity we come to intimate, personal objects that create an emotive connection between an individual and a collective movement: badges worn as public pledges of support, or jewellery made in prison that forms a link with campaigners on the outside. A final section of the exhibition is titled A Multitude of Struggles and comprises a series of stand-alone case studies. Here there is space for a more intense, less mediated engagement with the objects and the textures of personal narratives and design stories. The corresponding sections in this book illustrate some of the objects from the exhibition.

One of our principal methodological inspirations, besides the tradition of history from below, has been the use of participatory action research methods to engage with current movements. A fully developed action research approach wasn't possible, firstly, due to institutional constraints, and secondly, because we weren't engaging with a single local community context. But we aimed to be guided by the key values and principles of shaping research as a socially just activity: researching with, rather than on, communities; recognizing participants as experts and opening the research process to them; allowing them to fundamentally shape the research; and documenting outcomes in a way that is accessible and useful to those it claims to talk about. The research process of forming the exhibition's narrative and physical design has also been shaped by workshops with makers, movement participants and engaged academics. In the exhibition itself the objects are to be physically encountered as far as possible from the same perspectives as their users and makers, not upon pedestals or high upon a wall, yet resisting any attempt at a theatrical 'restaging', immersion or glib gallery participation. Normal V&A rules concerning plinths, barriers and touching distances have all been revised.

THE WORLD IS IN DIRE NEED OF A SOCIETY OF THE CREATIVELY MALADJUSTED. IT MAY WELL BE THAT THE SALVATION OF OUR WORLD LIES IN THE HANDS OF SUCH A CREATIVE MINORITY.

Martin Luther King, 'Transformed Nonconformism', *Strength to Love*, 1963

Foregrounding the ecology of these objects meant presenting as much context as possible (video, photographs, design notes, even hate mail), or highlighting process, for example by exhibiting the tools of clandestine solidarity street art in Syria – hidden stencils and disguised paint rather than the images on walls they produced. Sometimes supporting objects – for example, a letter written from prison – were more prominent and affecting than the object itself. Sometimes it made no sense to present the object itself at all outside of video, photograph or design diagram form. The exhibition is open to being actively read in different ways. Each object is accompanied not just by a curatorial text, but also by a statement from its users or makers, speaking in their own voice, given equal or higher visibility. The design diagrams in this volume, which take their lead from the way in which many movement how-to guides understand these objects, are also indicative of our approach to presenting individual objects. Such diagrams accompany them and are available to take away by the exhibition's exit. They offer another practical way to read the objects, inviting visitors to think tactically about why disobedient objects are made and the design problems they solve.

In caring for these objects, our attempts to engage critically with the Museum's organization in terms of sponsorship, education programmes, internships, 'late' events and the Museum shop have all become part of our project. Measuring its success will be difficult, not least in that any outcomes for the institution itself will be loudly projected, whereas those among movements will be quieter, less visible.

Like the objects it collects together, this project is incomplete. Its attempt to speak to the silences inevitably has to address its own, and it will fail in ways that we haven't yet anticipated.

Bringing these objects together in an art and design museum depended on personal trust, and proceeded through networks and connections that were necessarily limited in geographical and cultural terms. The histories it uncovers will be blighted by omissions of balance vis-à-vis the Western movements closest to us, incorrect precedents for designs and strategies, or gaps in genealogies. In attempting to attend to such limits, the exhibition and this book will launch alongside an online participatory archive, jumpstarted with their content. Further object histories can be recorded and their genealogies remixed and complicated. The exhibition itself only opens a crack. It isn't a complete synthesis or theoretical overview, but looks at some objects of disobedience through a series of particular, local moments tied in varying ways to movements for change. It is an invitation to look into the conditions and forms of making in social movements, to analyze omissions, connections, possibilities. More broadly, it might also evoke questions about how we define cultural value – and what art, design, arts education and museums are for.

BECAUSE REBELLION, FRIENDS AND ENEMIES, DOES NOT BELONG EXCLUSIVELY TO THE NEO-ZAPATISTAS. IT BELONGS TO HUMANITY. AND THAT IS SOMETHING THAT MUST BE CELEBRATED. EVERYWHERE, EVERYDAY AND ALL THE TIME. BECAUSE REBELLION IS ALSO A CELEBRATION.

Subcomandante Marcos,
EZLN communiqué, December 2013

barrels, chains, earth, manure, paving stones, carts, carriages, automobiles, buses, police cars, pallets, beams, planks, gates, windows, lamp posts, banisters, household furnishings, tables, bedsteads, dressers, doors, sinks, tyres, pianos, mouldings, armoires, urinals, dead horses, casements, stones, fences, wrought-iron, gunpowder, lampblack, pickaxes, drainpipes, saltpetre, crowbars, mattresses, wheeled platforms, blacksmith's anvils, sandbags, bullets—**Barricades as Material and Social Constructions**

The barricade has figured prominently in insurrectionary struggles for more than four centuries. The visibility, longevity, versatility and sporadic efficacy of this tactic of street warfare explain why it possesses a symbolic resonance that has made it a virtual icon of the revolutionary tradition. Yet the barricade constitutes, first and foremost, a concrete, tangible object: by definition, the term implies a physical structure, built and defended by citizen-insurgents, for the purpose of laying claim to urban space and challenging the constituted authorities.

Although I will later argue that an exclusive focus on the material properties of barricades can obscure their broader import, a grasp of the practical aspects of their construction and deployment is nonetheless a prerequisite for understanding the role they have played in deciding political conflicts and effecting social change in the modern era. This brief overview examines the raw materials used to build traditional barricades; the ways that those structures have been used in combat; and the forces that have reshaped barricades over the centuries, including those responsible for the changes in form and function that have occurred since the classic age of barricades ended in the mid-1800s.[1]

BUILDING BARRICADES

The physical make-up of these structures has determined how we conceive of them, beginning with the etymology of the word 'barricade' itself. The term derives from the Old French word *barrique*, referring to the hogsheads or barrels used by insurgents in the sixteenth century to reinforce the chains that rebellious local residents would stretch across strategic entry points to their neighbourhoods to bar access to the forces the government might send against them.[2] The need to swiftly improvise an effective form of protective cover prompted them to commandeer the casks that – because they were used for the transport and storage of wine, flour and other essential commodities – were everywhere to be found in early-modern cities. For insurgents, barrels presented several distinct advantages. First, when empty, they could be rolled on their sides to the desired location with little effort. Once in position, they could then be stood on end and filled with earth, manure or that other staple of traditional barricade construction, *pavés* or paving stones. These granite or sandstone blocks were even more ubiquitous than barrels, for urban street fighters of that day could be sure to find them right beneath their feet. Once aggregated and contained within an upright barrel, a heap of individual paving stones became an all but unmovable mass.

Right The first 'Day of the Barricades', Paris, 1588, with barrels prominently featured, from Louis-Pierre Anquetil's *Histoire de France* (1851).

We know from the testimony of eyewitness observers that barrels and paving stones were key components of structures raised in all of the major insurrectionary episodes of French history: in 1588 and 1648 as part of the first and second 'Day of the Barricades'; on Bastille Day in 1789 when a handful of barricades made a brief appearance; and during the July Days of 1830 (see p.28, right) and the February and June Days of 1848 (see p.29, top and bottom, respectively), the largest such events ever recorded as measured by the prevalence of barricades.[3] More surprising, perhaps, is the continuing use of these customary materials well into the twentieth century, despite the fact that the

state of urban warfare had changed dramatically. When, for example, student demonstrators in the May 1968 uprising in Paris adopted as their slogan '*sous les pavés, la plage*' ('beneath the cobblestones, the beach' or, more freely, 'liberation through insurrection'), this poetic turn of phrase (as well as the adoption of tried-and-true techniques of barricade construction) was a way of connecting their insurrectionary urge to the rich revolutionary heritage of the French capital.

Yet even when modern conditions prompted changes in construction techniques, barricade builders consistently sought to retain or replicate the advantages that their original materials had possessed. Barrels may have become more scarce in nineteenth-century cities, but the use of wheeled vehicles had grown apace, and carts, carriages and wagons had many of the same attributes to recommend them: they could be quickly rolled into position and yet, once turned on their sides and wedged in place with paving stones and other scavenged materials, would resist the most dogged efforts to dislodge them (see below, and opposite top). In more recent times, automobiles, buses and – most provocatively – police cars have been pressed into service in much the same way. What these and the previous examples tell us is that the most salient properties of the objects from which barricades have been fashioned are their ready availability, their selective mobility and their facility for being creatively combined into an unyielding mass.

Beyond these essential qualities, the constituent elements of conventional barricades are most remarkable for their sheer diversity. Insurgents would scour nearby construction sites in search of beams or planks that could add solidity to the emerging structure. They would appropriate the gates and fences of public gardens, the metal grates from the base of trees that lined the streets, the trees themselves, lamp posts and even the wrought-iron banisters of interior stairways, for such components collectively formed a basis around which looser materials could be knitted together. Sympathetic neighbours (or those who could be easily intimidated) might donate household furnishings, often thrown from upper-storey windows into the streets below. All such contributions were welcome: chairs and tables, bedsteads and dressers, doors and mouldings, even

Top This contemporaneous depiction of the second 'Day of the Barricades' (Paris, 1648), in which both barrels and chains are in evidence, is the oldest known representation of a barricade.

Right Barricade construction in July 1830, from Armand Dayot's *Journées révolutionnaires, 1830, 1848: D'après les peintures, sculptures, dessins . . . du temps.*

Opposite Top A barricade from the 'February Days' of 1848, rue Saint-Martin, Paris, *Illustrated London News*, 4 March 1848.

Opposite Bottom The Great Barricade at the entrance of the Faubourg Saint-Antoine, Paris, June 1848, *Illustrated London News*, 1 July 1848.

the occasional armoire or sink. They sometimes found a place alongside the more exotic items mentioned in contemporary sources: a piano, a blacksmith's anvil, public urinals, or in one case, the body of a dead horse.[4]

Barricade construction was essentially a spontaneous rather than a studied process and one that took place under acute pressure of time and the threat of imminent, violent confrontation. The basic structure of most barricades was completed within a two-hour span (and in some cases as little as 15 minutes). Insurgents might subsequently reinforce or repair their creations as the vagaries of battle permitted, and they often took advantage of nightfall to reoccupy or rebuild barricades lost in the previous day's fighting. Yet typical barricades, despite being hurriedly constructed from found materials, exhibited a remarkable consistency over the first 300 years of the technique's existence.

THE DEPLOYMENT OF BARRICADES

The high degree of effectiveness that barricades sometimes demonstrated must be attributed as much to the way that they were utilized as to the physical components of which they were made. The high degree of continuity in the selection of barricade sites shows us that their deployment was by no means haphazard. The revolutions of July 1830 and February 1848 involved massive and successful insurrections that were so meticulously documented that we possess detailed maps showing the precise locations of thousands of individual barricades. By focusing on the specific neighbourhood surrounding the Eglise Saint-Merry in Paris, a traditional hotbed of insurgent activity, my research has shown that more than half of all the 1848 barricades in that vicinity were built on the identical spot where just such a structure had stood in 1830.[5]

The choice of emplacement was frequently made by insurgents who had been thrust into a leadership role by virtue of their prior military experience, however humble. Even a rudimentary grasp of tactics was usually sufficient to identify the advantages of certain sites, so it was no accident that the archetypal barricade was located in a narrow, winding street in one of the oldest and most dilapidated quarters of the city. Such a venue allowed insurgents to quickly block circulation,

after which they might build their structure up to a height that at a minimum afforded protection from attack and that sometimes reached as high as second-storey balconies. Constricted quarters, twisting angles and limited lines of sight also inhibited the military authorities' use of their most effective weapons, cavalry and artillery. In addition, these environs usually offered insurgents the most favourable prospects for recruiting reinforcements from a sympathetic population. Still, the insurrectionary history of Paris, where these sorts of events occurred most frequently,[6] shows us that, notwithstanding all the supposed benefits that accrued from building in confined spaces in working-class quarters, nineteenth-century barricades might well be found almost anywhere: in wide-open public squares, on or adjoining the many bridges across the Seine, in largely symbolic locations like the vacant former site of the Bastille and (at least in very large uprisings that mobilized virtually the entire Parisian population) in well-to-do neighbourhoods in the western districts of the capital.

In assessing its utility, the typical barricade should never be thought of as a stand-alone feature of an insurrectionary landscape, for it was most often tightly integrated with adjacent residential buildings or commercial establishments. Insurgents might set up their headquarters in the corner wine shop or might rededicate a neighbourhood pharmacy as a first-aid station. A nearby workshop might be converted into a tiny munitions factory where lead from casement windows and drainpipes could be melted down and recast into bullets, while gunpowder was fabricated from lampblack and saltpetre, scraped from damp cellar walls. Even more critical were the strategic advantages to be gained by integrating barricades with adjacent apartment buildings. By posting snipers at their windows, insurgents could set up deadly crossfires and inflict heavy casualties on attackers. Using pickaxes and crowbars, they could open gaping holes in exterior walls that allowed them access to basement passageways and thus a ready avenue of escape in case the barricade should fall.

It was common, moreover, for barricades to be used in combination rather than singly. Charles Jeanne was commander of three contiguous barricades near the Eglise Saint-Merry as well as the several shops and apartment houses that his men also controlled on 6 June 1832. His account of the fighting makes it clear that insurgents were able to hold out against the initial fierce attacks that army and National Guard units launched against them thanks only to the savvy use of an urban environment that they, as local residents, knew intimately.[7] Other contemporary sources

'Rise like Lions after slumber
In unvanquishable number,
Shake your chains to earth like dew
Which in sleep had fallen on you –
Ye are many – they are few.' Percy Bysshe
Shelley, *The Masque of Anarchy*, 1819

that document the use of multiple barricades provide revealing insights into the constant back-and-forth between insurgents and social control forces. In an attempt to counteract the firepower of the army, two or more parallel barricades might be spaced at 50- or 100-yard intervals.[8] Insurgents would defend their forward position until the barricade had been largely demolished by enemy artillery. They would then fall back through narrow passageways left at either end of the barricade to their rear and thus be able to maintain a harassing fire against the ongoing assault. The artillerymen, meanwhile, would have great difficulty advancing their field guns across the rubble-strewn surface that the insurgents had left behind.

The need to offset the devastating effect of cannon was also responsible for insurgents' preoccupation with improving the form of the barricades they built. Auguste Blanqui, the consummate revolutionary of the mid-nineteenth century, actually drew up plans for an ideal barricade, complete with diagrams and elaborate calculations of the length of pavement that needed to be dug up in order to furnish the required number of paving stones. His illustration included a *glacis* or sloped embankment, erected right in front of the barricade proper to protect it from artillery shells.[9] His treatise was never published in his lifetime and his project was never implemented in an actual uprising, but by the middle of the nineteenth century insurgents were already devising more practical alternatives involving, for example, barricades constructed in V shapes, their points aimed directly at attacking forces. The effect of this innovative profile was to deflect cannonballs that struck the face of the barricade obliquely and to absorb the force of those that landed head-on in such a way that the barricade collapsed in upon itself, thus retaining its integrity.[10]

The history of the barricade has been marked by a constant struggle for superiority, involving innovation and mutual adaptation on the part of the hostile camps in civil conflicts. For example, the June Days of 1848 were barely over before the army began receiving proposals for the construction of mobile barricades, touted as a means of 'preventing or attenuating losses in the future'.[11] Two decades later, a working prototype was actually placed

in service, but this time its champions were the insurgents of the Paris Commune. Their variant, which consisted of mattresses packed with stones and mounted on a wheeled platform, was just one of several innovations the Communards tried out. These experiments ranged from the introduction of new materials (such as barricades constructed almost entirely of sandbags) to novel forms of organization (like the official Barricade Commission that oversaw construction activity city-wide). The same process of give-and-take that had led Napoleon to employ 'a whiff of grapeshot' to prevent the construction of a barricade during the *journée*, or revolutionary day, of Vendémiaire in October 1795 would ultimately give us the tactical use of

Opposite Barricade in an open square, Paris, February 1848, *Journées illustrées de la révolution de 1848: Récit historique de tous les événements … accompagné de 600 gravures. 1848–49.*

Right Artist's conception of a mobile barricade, 1848 • *Journées illustrées de la révolution de 1848: Récit historique de tous les événements … accompagné de 600 gravures. 1848–49.*

rubber bullets and tear gas (sometimes delivered by helicopters) as the means of overpowering barricades in more recent times.[12] Meanwhile, in the insurgent camp, the drive to innovate would yield such present-day inventions as barricades made from burning tyres or so-called 'intelligent barricades' (manned, prefabricated tripods) that selectively bar motorized vehicles from inner-city streets while permitting free access to pedestrians.[13]

THE CHANGING TERMS OF BARRICADE COMBAT

We must not allow the present focus on the material and pragmatic aspects of barricade combat to overshadow the fundamental truth that armed insurrection is ultimately a moral struggle in which the barricade fulfils a variety of functions, among which the cultural and symbolic have, over time, become paramount. Keep in mind, as a useful rule of thumb, that insurrections undertaken by irregular civilian forces against better trained, equipped and organized soldiers and police consistently fail unless they manage to win over a significant segment of their opponents, thus destroying the government's claim to legitimacy and the morale of its supporters.[14] For this reason, the value of the barricade lies mainly in its ability to bring insurgents and repressors face to face, at close quarters, making possible the crucial interactions without which the rebels have little hope of success.

I do not wish to understate the purely pragmatic benefits that street fighters have historically derived from their use of barricades, particularly between the time of their origination in the sixteenth century and their heyday in the first half of the nineteenth. However, much of that practical efficacy was gradually lost in the face of: 1) new technologies and, just as important, changing norms regarding when and how it was appropriate to use these innovations against the domestic population;[15] 2) more sophisticated planning on the part of military authorities, who developed systematic strategies for the suppression of civil disorders;[16] and 3) a broadening of the alternatives available to ordinary people for expressing and enacting their desire for change – most notably, organized political parties and social movements – thus diminishing the allure of violent revolution.

And yet, despite all this, the barricade did not disappear! It has, in fact, not just survived in our day of instant communication and electronic warfare but actually flourished, often in entirely new settings. It owes this persistence and vitality to its capacity to operate not simply as a way of seeking refuge from physical assault but, more abstractly, as a means of claiming one's place in a revolutionary lineage, even as one simultaneously declares allegiance to a cause and seeks to recruit others of like mind to join the struggle.

In our day, building a barricade, however varied its outward form, is a symbolic gesture whose meaning cannot be separated from the historic events of the classic age of revolution and the world-shattering consequences they engendered. The genius of the barricade was to have taken malleable materials and transformed them into an always recognizable but ever evolving instrument of popular insurrection. It may have begun as a uniquely French invention and remained exclusive to that country through most of the eighteenth century, but by the mid-nineteenth it had spread across the Continent, and in the twentieth had become truly global in its reach. This remarkably rapid and wide-ranging diffusion reminds us that the most effective techniques of social movement mobilization are those that, like the barricade, combine simplicity and adaptability.

Opposite Top The Great Barricade at the entrance of the Faubourg Saint-Antoine, Paris, June 1848, *Illustrated London News*, 1 July 1848.

Opposite Bottom Protesters stand on a barricade during #occupygezi protests in Istanbul, 3 June 2013. The protests began as a peaceful occupation to protect a park from demolition, but after violent repression they grew into national demonstrations against the government. Barricades were a new tactic in these protests, built in concentric rings in streets around occupied areas. They were also built with a self-conscious sense of historical symbolism and cultural importance. Some were named after fallen protesters, such as Abdullah Cömert; some became sites for meetings and performances (one barricade was 'played' by a team of drummers); while others were continually reconstructed (every brick of one being decorated with its own smiley face).

COMPLET

books, kettles, sleeping bags, cooking burners, locks, food, toilets, media devices, ladders, metal fences, spanners, shields, mobile phones, cooking grills, maps, shopping trolleys, lock-ons, rope, gaffer tape, exhaust vents, tripods, buses, wood scraps, sandbags, trucks, gas masks, barricades, slingshots, caravans, tarpaulins, teacups, shields, propane heaters, rain ponchos, cardboard boxes, pop-up tents, computer equipment, power chargers—**The Disobedient Objects of Protest Camps**

At just before midnight on 9 December 2011 in Foley Square, Lower Manhattan, dozens of protesters amassed on NBC's set for the television series *Law & Order: Special Victims Unit*. The square had been re-designed as a mock Zuccotti Park, home of New York City's Occupy Wall Street encampment. Built just three weeks after the camp's eviction, the set resembled the site in its prime – dotted with hand-painted signs and banners, and with spaces designated for the 'People's Kitchen' and 'People's Library'.

Among other online and offline organizing tools, protesters used Twitter with the hashtag 'mockupy' to gather people together for the action. 'Mockupiers' noted in their mobilizing messages that it was easier to get a permit to hold a fake protest than it was to get one for the real thing in New York City. At around 11.30pm the mockupiers crashed the set, flipping through pages of the library's books and eating food from the kitchen. Messages were marked on to posters, chanted into television news cameras and recorded on protesters' own mobile phones and live-stream feeds. These included 'We are a movement, not a TV plot' and 'Occupy Wall Street is not for sale'. The mockupation of the *Law & Order* set continued until 50 police turned up, revoking the television crew's permit and telling the protesters to leave or face arrest.

'Disobedience, to be civil, implies discipline, thought, care, attention.'
Mahatma Gandhi, attributed, 1920s

Unlike marches, strikes or demos, protest camps are unique as place-based sites of on-going protest and daily social acts of 're-creation'.[1] The protest camp is a home-place.[2] Its sleeping shelters, kitchens, bathrooms, meeting spaces and often library, crèche, on-site toilets and well-being spaces distinguish the camp spatially and temporally from other social movement sites. The protest camp is rich with infrastructures and objects, from media devices to makeshift cooking burners to the ladders, spanners and locks of direct action toolkits. *Law & Order*'s set designers were well aware of this, mimicking its organizational framework and recasting its individual elements – from books to tea kettles – as props, in effect materializing a functional world of protest.

WHOSE FAKE PARK?

At Mockupy, protesters challenged the mainstream media's appropriation of their movements' iconography, objects and distinctive architectures by disrupting, or in Guy Debord's terms, 'detourning' the *Law & Order* set; an act of reclamation that was echoed in the Mockupy chant 'Whose fake park? Our fake park.' Yet there is something more at stake in the declaration 'Our fake park'. The mockupation of Foley Square also points to another phenomenon at work – the power of the object world to propel us into action. By recreating the objects of protest camps as part of the set design, the fake park drew people into the encampment. There were books to read, kettles to be boiled, picket signs to carry. Mockupy shows the vibrancy of the non-human world. It speaks to us of what objects can compel the body to do, calling for touch, for care, for action, drawing us into their world.[3]

One way of thinking about objects is in terms of what Ian Hodder calls human-nonhuman co-dependencies or 'entanglements'.[4] When things are entangled together they are caught up, twisted, difficult to untether. For Hodder, objects are bound in co-dependencies, both with humans and with other non-humans, in what Ruth Cowan has referred to as 'technological chains'.[5] In other words, we might think of our smartphones as tools, objects that allow us to do all sorts of things. But just as we 'depend' on our smartphones to send a text or give us directions, our smartphones also 'depend' on us. In order for them to work as our tools we have to keep them charged, connected and ready for use. To protect them, we might buy a rubber crash-proof casing or leather flip-top.

Taking this idea of object co-dependencies into the protest camp, we can start to think about declarations of 'Twitter revolutions' and 'Facebook uprisings'. Yes, Twitter and Facebook play a key role in mobilizing contemporary protest campers. The

rise of social media has sped up the protest cycle, turning both news media and activist media-making into 24/7 tasks. But for social media to work as a tool for activists they need both content and power. Beyond the 'human microphone' most communications technologies depend on electricity. As researcher Patrick McCurdy found out, it is the street lamps, external hard drives and nearby cafes that make it possible to stay connected in an outdoor environment – especially during service-provider blackouts like those experienced in Cairo's Tahrir Square from 2011. Weather also affects the ability to engage communication technologies on a campsite – blizzards and thunderstorms can make working with electronic devices quite a challenge. But at protest camps, people find innovative ways to get power for their messages. Around Tahrir Square, protesters who had run out of charge would tap into the wires of street lamps, re-routing electricity, a practice popularized in favelas and squats. During the Ukraine's Orange Revolution (2004), the residents of the mass protest camps that emerged in opposition to electoral corruption used so much power that new telephone poles had to be installed.[6] At Occupy Wall Street a system was developed whereby if it looked like rain, media team members would wrap up computer equipment like presents in tarpaulins and donated rain ponchos, placing them carefully into a large skip also lined with waterproofs. As Occupy spread to new cities and new climates, protest campers in Occupy Anchorage, Alaska, faced severe cold. When temperatures went down too low to use a computer, live streaming had to be stopped. In an effort to keep filming, inventive occupiers brought in propane heaters to keep the computer at operating temperature. The computers, phones and power chargers of protest camps are incredible tools in the hands of protesters. But just as they serve people's desires to communicate, people get called upon to serve them, their power often relying on protesters' ingenuity.

While they may be co-dependent, the objects of protest camps are not neutral. A teacup in your kitchen offered to a house guest is not the same as a teacup in a protest camp kitchen handed to a stranger-turned-comrade behind a barricade. The bottle of vinegar in your cupboard for dressing the chips is not the same as the bottle of vinegar stocked in the makeshift field hospital of an occupied square for rinsing out tear-gassed eyes. Objects acquire meaning in their assemblages, or the ways in which they are arranged with other

technologies, bodies and environments. While the cup or bottle in my cupboard may not be 'disobedient', when repurposed and placed into the protest camp as objects of affinity and counter-repression, they become 'disobedient'.

Other protest camp objects are designed, we might say, for disobedience. Shields adorned with images of climate refugees, used to transport pop-up tents at London's Heathrow Climate Camp

Opposite Left Painted map from Acampada Sol with committee locations • Gouache, marker and pen on white industrial paper • Madrid, 2011 • M15 Archive • The Acampada Sol in Puerto del Sol emerged from the M15 or Indignados movement.

Opposite Right Map of Kingsnorth Camp for Climate Action, Kingsnorth, Britain, 2008.

Top Shield at Heathrow Climate Camp, London, 2007, depicting photographic portraits of people seriously affected by climate change. Each shield – in effect transformed cardboard packaging – contained a pop-up tent used shortly afterwards in the blockading of the entrance to the British Airports Authority offices.

(2007), are an excellent example. With gaffer tape, rope and a bit of ingenuity, protesters transformed the square cardboard boxes packaging the tents into vibrant symbolic shields that led the march to BAA headquarters. Similarly, the concrete lock-ons (obstruction tools sometimes known as 'dragons'), made infamous in the anti-roads protests of the 1990s, show how protest camps can become R&D sites of disobedient design innovation. As the Road Raging strategy handbook encouraged its readers:

> Direct action is an evolving art form – 'Necessity breeds ingenuity'. Remember that the enemy have avidly read this and every other similar guide, and will be constantly devising methods to beat the 'tools' described – so you MUST innovate, improve and invent. Your imagination is the limit! Various different methods of obstruction can be used in combination.[7]

DISOBEDIENT TENTS

Perhaps the most prominent disobedient object of protest camps is the tent. Tents are disobedient in a number of ways but most basically, perhaps, when acting as signboards. Whether affixed with banners and posters, or painted and drawn on, the surfaces of tents bear the messages of a movement, its multiple perspectives and participants. In this way tents act akin to more traditional political signs, directing us to vote or donate to a campaign. At protest camps visual signs often articulate what Fabian Frenzel describes as a politics of exceptionality,[8] using language and images that call into question the architectures of the 'settled' nation-state.[9]

Such signs often challenge and re-imagine land rights and border controls. The American Indian Movement's encampment at Occupy Alcatraz (1969) declared 'This is Indian Land' and, a few years later, Australian aboriginal activists stuck a sign in the lawn of Parliament House, Canberra, proclaiming their own 'Aboriginal Tent Embassy' as a means of confronting unjust laws that contravened aboriginal land rights. In 1970s Germany, anti-nuclear protesters declared themselves a 'Free Republic of Wendland' and in the United States, the 'Minnehaha Free State' was born in a 1998 anti-road building protest camp. More recently, the 2007 Heathrow Climate Camp offered a walk-through airplane door reconstruction that invited campers to 'Exit the System' and in 2011, at the Occupy London encampment at the foot of St Paul's Cathedral, a tent proclaimed itself the 'Former Soviet Republic of Tentistan' where 'All British Law is Null and Void'.

The materiality of the tent enables these performative encounters of land reclamation. Physically occupying – and often squatting – space, protest camp tents are not merely the props of an elaborate site-specific theatre performance. As architectural structures, they themselves occupy space, claiming their site of erection as home. Yet unlike recreational camping, the challenge protest camps pose often lies in their refusal to 'go home'. It is this disobedience that moves tents from symbols of resistance to objects of defiance.

In the United States courts have repeatedly upheld the status of tents as a form of protected symbolic communication. However, when tents move from being symbols of protest to being usable infrastructures for sleeping, eating and other forms of recreation, prosecutors trot out 'reasonable time and place restrictions'. For example, when the Occupy protest movement spread coast-to-coast reaching Fort Myers, Florida, an ensuing trial deemed that 'fake sleeping' was an acceptable mode of protest, while real sleeping was outlawed. The court ruled that it was in the city's interests to close the park at night and prohibit 'protracted lounging' at all hours.[10] In other words, so long as tents function solely as objects to be looked at, they can be permitted. However, when the structure moves from being a piece of art to offering actual shelter, it begins to be seen as a threat to public

'If you go to one demonstration and then go home, that's something, but the people in power can live with that. What they can't live with is sustained pressure that keeps building, organizations that keep doing things, people that keep learning lessons from the last time and doing it better the next time.'
Noam Chomsky, *What Uncle Sam Really Wants*, 1993

order. This view was made explicit following the ruling to evict the original Occupy London camp, located outside the Stock Exchange. Liberal Democrat MEP for London, Sarah Ludford, commented on the verdict: 'Protests should not morph into tent cities. The right to protest is too precious to be undermined by long-term encampments which disrupt normal life to an unacceptable extent, beyond the inevitable and legitimate inconvenience of a one-off demo.'[11]

A similar strategy for repressing protest camps was used in the Parliament Square Peace Campaign, started by campaigner Brian Haw on 2 June 2001 in front of London's Westminster Palace. After almost a decade of legal battles with, and appeals to, local and national government, the Police Reform and Social Responsibility Bill came out banning 'any tent or any other structure that is designed, or adapted … for the purpose of facilitating sleeping or staying in' and received Royal Assent on 15 September 2011.[12]

Architect Greg Cowan argues that tents and nomadic camping practices have a long history of disobedience, challenging the idea that established or settled architectures should dominate how we live. This is why travellers, caravans, carnivals and other mobile ways of life and work are often seen as threatening. Influenced by avant-garde artistic practices, in the 1960s some protesters in the West began to adopt tent-like tactics. Groups such as the San Francisco Diggers, followed in the 1970s by Britain's Artists Placement Group, took on nomadic and camping practices, setting up mobile food stations and occupying public parks. Mobile and inflatable architectural design, as used by Italy's UFO Group, entered the topography of demonstrations and occupations.

Akin to these influences in the 1960s and 1970s, in October 2011 Greta Hansen, Kyung-Jae Kim, Andy Rauchut and Adam Koogler came together as 123 Occupy to build strategies for occupation. Their work combined architectural structures, urban design principles and an open-source ethos with a commitment to community-building, inspired by the Occupy movement and radical design. Influences included architectural artist Michael Rakowitz's parasite project, in which he created a series of blow-up pods for the homeless that

Opposite A tripod in use at the 2007 Heathrow Climate Camp.

Right Working diagrams for scaffolding tripod designs (detail, see p.64 for full reproduction), drawn by B. Dahl • Pen, Tippex and glue on card • February 1997 • Private collection.

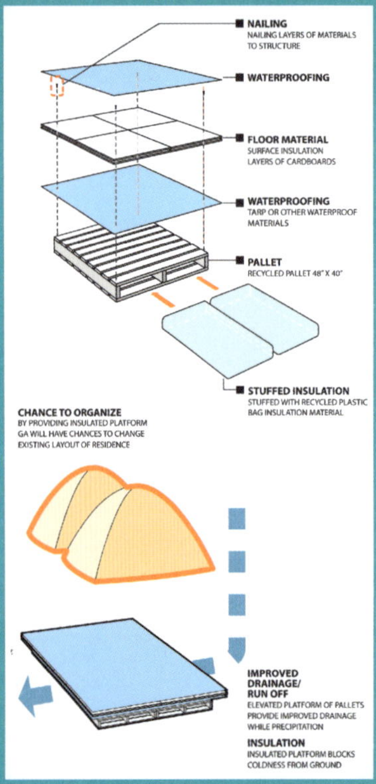

Top Working drawings and prototypes for an insulated pallet, 123 Occupy • Wooden pallet, cardboard, plastic bags and tarpaulin • New York, 2011.

Opposite Members of Greenham Common Women's Peace Camp blockading the entrance to RAF Brawdy, Britain, 1982. Greenham Common camp was established in Berkshire in 1981 to challenge the siting of American nuclear cruise missiles at an RAF base there. The camp quickly became a focus for imaginative direct action. The following year the women marched from Cardiff to RAF Brawdy in Pembrokeshire to protest against US Army underwater sound surveillance activity. The banner pictured was made by Thalia Campbell.

borrowed warm air from the exhaust vents of buildings. As Hansen explained:

> We started working together when the occupation of Zuccotti Park was occupying everyone's attention. The first thought was 'what are they going to do in the winter?' And the second thought was opportunistic. As architects we thought we could try to engage what we already cared about with what was an open ground for ideas. We worked with the Architecture Working Group to brainstorm solutions for the winter, and then we developed a raised insulated water-resistant platform for tents and a collapsible triage station for the medical pavilion.[13]

Such tents carry designed qualities of disobedience that become visible when they are made explicit in acts of protest. When tents move from being temporary, ephemeral structures – for sleeping in at a festival, for example, or being set up on the street as part of a short-lived protest – to on-going shelters, their disobedience is magnified. Barbara Ehrenreich drew attention to this in her commentary on the Occupy movement: 'What the Occupy Wall Streeters are beginning to discover, and homeless people have known all along, is that most ordinary, biologically necessary activities are illegal when performed in American streets – not just peeing, but sitting, lying down, and sleeping.'[14] In this sense these tents tell us about the inadequacies of our governments to provide for their populations. They reveal the realities of homelessness and the everyday rules that restrict where and when we can sit, eat and sleep.

Some protest camps take acts of autonomy and challenges to the state further, erecting barricades as a means of protection and empowerment. In 2006 in Oaxaca, Mexico, a teacher's sit-in turned into a much larger struggle as people took over entire streets, squares and neighbourhoods in protest. Barricades were erected out of metal and wood scraps and sandbags, as well as buses and trucks. Some barricades went up to secure the people's radio stations and occupy media outlets. At one cultural event, films of the uprising were screened on the side of a barricade bus. Barricades were places of defiance and vulnerability, the site of many police attacks and collective resistance. Inventions designed for barricade defence included homemade gas masks, slingshots and a communication warning system using fireworks.[15]

DISOBEDIENT KITCHENS

At protest camps one of the key places to meet and eat is the kitchen. In stark contrast to the traditional white marble and whisky club spaces of political arenas, the protest camp kitchen is often explicitly rendered a meeting space and site of governance; it serves as a locus of communicative interaction. Likewise, whereas a dining-room table signifies the forum of men's deliberations apart from the labours of housework, the kitchen table marks the space of conversations that occur during collective, domestic work. This is seen in the 'barrio', or neighbourhood city planning, of the World Social Forum youth camps in Brazil's Porto Alegre[16] as well as the No Border camps for migrant justice and freedom of movement, which began in 2002 and spread throughout Europe, and at climate camps in Britain.

Neighbourhood structures that centred around the kitchen as a central hub had been seen earlier in Britain's Greenham Common Women's Peace Camp, constructed in protest against nuclear armament in the 1980s, and at the anti-roads camps of the 1990s. Describing life at a British anti-roads camp in 1996, a member of the Dongas tribe (a

group of road protesters and travellers that grew out of the Twyford Down campaign) explained that aside from her treehouse bedroom, 'The rest of my home, it's basically the fire. That's where everybody meets around for our evening meals. That's where everything is centred. It's the dining room and the kitchen and the living room.'[17]

As a place in which to serve food, in addition to acting as a space for the exchange of information and objects, the protest camp kitchen becomes a high traffic zone. Documenting the 2007 anti-G8 protests in Germany, cartoonist and kitchen coordinator Morgenmuffel discusses the many odd requests that come to the kitchen during action planning, as people ask for various objects and information.[18] Similarly, protesters at Occupy Montreal told of how the kitchen began to serve as a de facto media tent, where whoever was on hand would act as spokesperson when approached by the media. One Montreal occupier explained that this often happened to be the more organized campers: 'Another reason why it came so much to the kitchen [was] because the media would always come first thing in the morning, and basically, the only people who are up at that time are at the logistical end of the camp.'[19]

Describing how these domestic spaces become political at a protest camp, Nick Couldry has written that Greenham Common 'turned inside out' the 'regular patterns' separating domestic/non-domestic, public/non-public and mediated/non-mediated space.[20] Greenham women also performed material culture-jams of kitchen life. They turned parts of the military base fence into a cooking grill and used hacksawed off shopping trolley wheels to make mobile kitchen storage for evictions. Morgenmuffel also documents re-design practices for shopping trolleys that occurred at an anti-G8 camp at the Hori-Zone eco-village in Gleneagles, Scotland (2005): 'Nearly as useful as gaffa [*sic*] tape ... during the G8, resourceful protesters displayed the many uses of the humble shopping trolley ... Transport for Kitchens, Veg Wash, Instant Barricade, Barbeques, Battering Ram, Brick transport and collection, Kids entertainment.'[21] Whether a journalist interviews the cook, or a shopping trolley is turned into a barricade, protest camps are spaces where architectures and objects are re-imagined and re-designed to build new political possibilities.

ARCHAEOLOGIES OF RESISTANCE

More than a set or a prop, it is the disobedient objects that make protest camps so unique among forms of activism. These brief sketches show how protest camp objects each have their own stories to tell. Some of these stories are painted on the canvas walls of tents, others lurk in cups of tea strewn around a campfire and yet others must be excavated from the tear-gas canisters that line the squares and streets after a camp's eviction. However, it is not enough to construct or define protest as a monument. It is not enough to collect social movement materials in a museum under a glass case. We need to find new ways to excavate and archive resistance, ways that do not erase their sense of place, of context, of the means by which they are shared across cities and countrysides alike. To build movement histories that can challenge the structures of power, we need what Yvonne Marshall calls 'archaeologies of resistance': 'It is in the juxtaposition of object, place and story that meaning resides – each makes the other meaningful and evocative. If the objects are separated from their context and lose their stories they become just another piece of plastic, metal, or concrete.'[22] An archaeology of resistance invites us to listen to these objects, to discover their stories, to remember that we depend on them just as they depend on us. Learning from protest campers' unruly entanglements with objects, we can continue to craft disobedience together.

Opposite 'The Feeding of the Five Thousand' (detail), design by Isy Morgenmuffel • Print on paper • Published in *Shut Them Down! The G8, Gleneagles 2005 and the Movement of Movements*, 2006.

nearly as useful as gaffa tape...
the SHOPPING TROLLEY!
DURING THE G8, RESOURCEFUL PROTESTORS DISPLAYED THE MANY USES OF THE HUMBLE SHOPPING TROLLEY...
TRANSPORT FOR KITCHENS
CAREFUL IT'S BUMPY HERE
CAN'T SEE ANYTHING ANYWAY
KIDS' ENTERTAINMENT
YAY!
VEG WASH
OI! GET THE SPUDS, NOT ME!
BRICK TRANSPORT!..
ROCKS! ANYONE FOR ROCKS?!
INSTANT BARRICADE!
..AND COLLECTION
WE MAKE RE-CYCLING, JA?
SEATING
BARBEQUE
MMMH..
YIKES!
BATTERING RAM
AAAAAH
(bottom bott-croppered off)
"A kingdom for a trolley!"

'We are forming the structure of the new society within the shell of the old.'
Preamble to the International Workers of the World Constitution, 1905

Mak-ing Worlds

Previous Traditional Turkish dervish dancer wearing a gas mask, Istanbul, 2013.

Right Village sign, produced by the Karnataka State Farmers' Association (KRRS), Mogenahalli, India, 1982. The KRRS has about ten million members, who use direct action against the neo-liberal developments that are affecting farming in India. Its aim is that of a 'village republic', with direct democracy and autonomy operating from the level of the village unit upwards. It was the first association to target the World Trade Organization and the issue of corporate globalization, setting up huge demonstrations. In October 1982 a village sign like this was put up to increase recognition of farmers, living in a context in which, for example, they were not offered seats when meeting state officials. The board reads: 'Officials and policemen, you may enter the village, on condition you are polite and civil. Farmers are available to meet between 6 and 7pm, and on Mondays.' The simple sign initiated a new arrangement, and versions of the board are still used at the entrance to many villages. The movement has been at the forefront of many important social changes in India.

Bottom Photograph and diagram by #occupygezi Architecture, a group of architects who documented the temporary structures built during the 2013 Istanbul protests.

Opposite A page from an anonymous pamphlet, *How to Protest Intelligently*, distributed in Tahrir Square, Cairo, 2011. An identical pamphlet in Ukrainian appeared in Independence Square, Kiev, 2014. It seems to have been translated online by protesters in late 2013, but Russian media have suggested it was produced by an NGO such as CANVAS. This and similar organizations have been accused by many of fostering dissent abroad in the interests of US foreign policy.

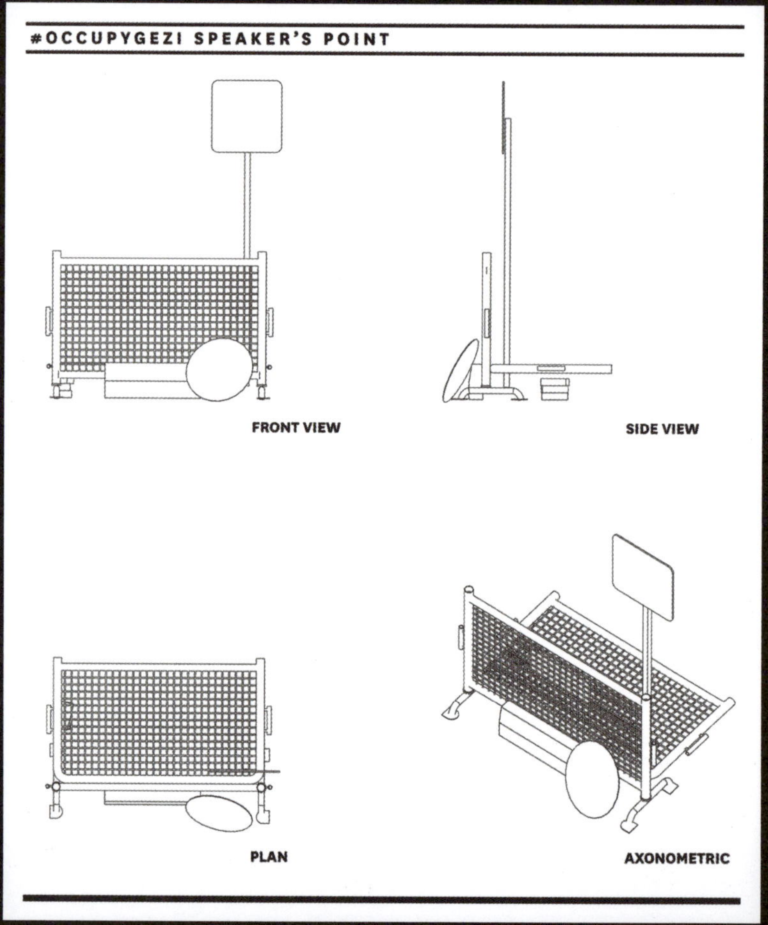

ملابس و أدوات ضرورية

سويت شيرت أو سويتر (أبو زعبوط)، فهو يساعد على إبعاد غازات القنابل المسيلة للدموع عن وجهك.

نظارة واقية (يمكن شرائها من أي محل حدايد وبويات)

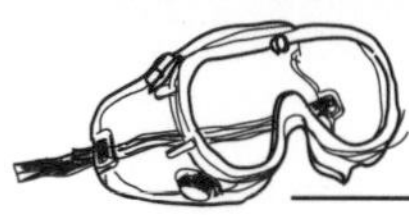

غطاء حلة، يمكن إستخدامها كدرع ضد ضربات الأمن المركزي بالعصا أو الرصاص المطاطي.

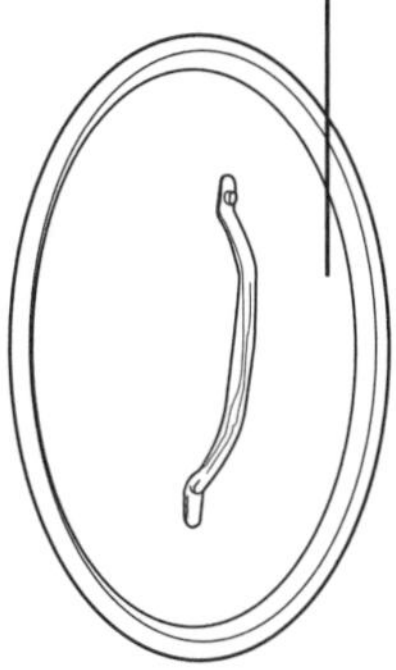

كوفية لحماية فمك ورئتيك من الغازات المسيلة للدموع

وردة، كي نعمل اللي علينا ونبدأ بتجمهر فى منتهى السلمية.

دوكو رش، علشان لو حدث ضرب من قبل السلطات، نرش الدوكو على زجاج الخوز والمدرعات لحجب رؤيتهم وشل حركتهم

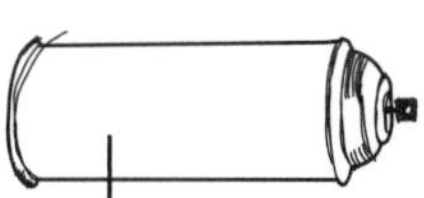

جوانتيات محارة، تساعد على حماية يداك من حرارة القنابل المسيلة للدموع

حذاء مريح للجري والحركة السريعة

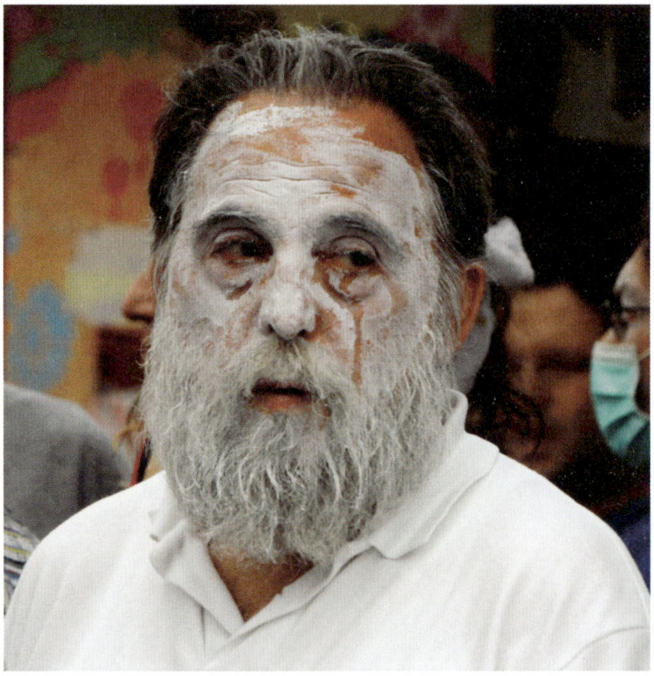

Top The Turkish government used record amounts of tear gas to disperse the 2013 Istanbul protests. Protesters devised homemade gas masks as a form of protection.

Right In 2013, during huge anti-austerity protests in Greece, demonstrators found a 50/50 solution of liquid antacid (Maalox) and water, sprayed on to the face, offered relief from the effects of tear gas, but left a white residue that marked protesters out to police. Pocket-sized sachets of antacid in the form of an oral gel (Riopan), which left no residue, became the popular remedy.

MAKESHIFT TEAR-GAS MASK

1

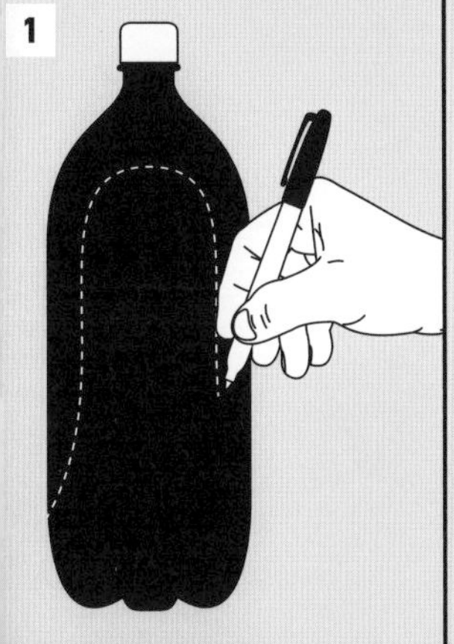

Use the permanent marker to draw a U-shaped area big enough to fit your face.

2

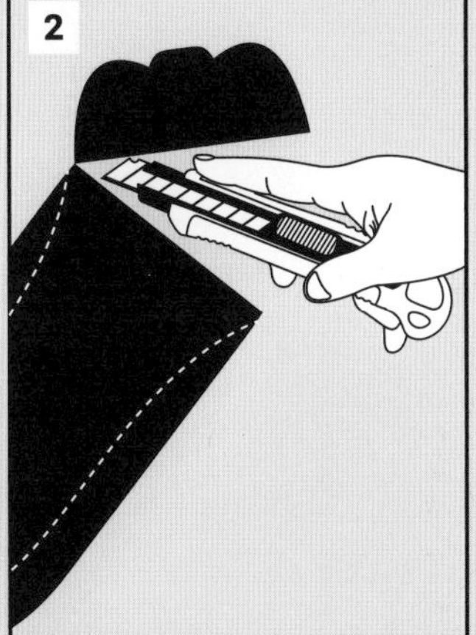

Cut away the bottom of the bottle just above the ridged area and discard it.

3

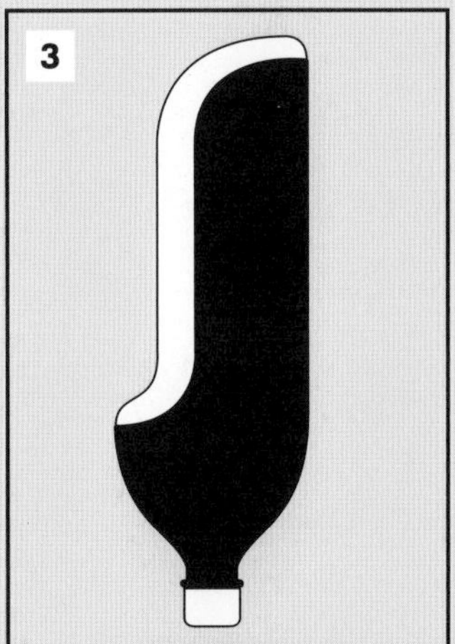

Cut along the lines of the template to remove the U-shaped section.

4

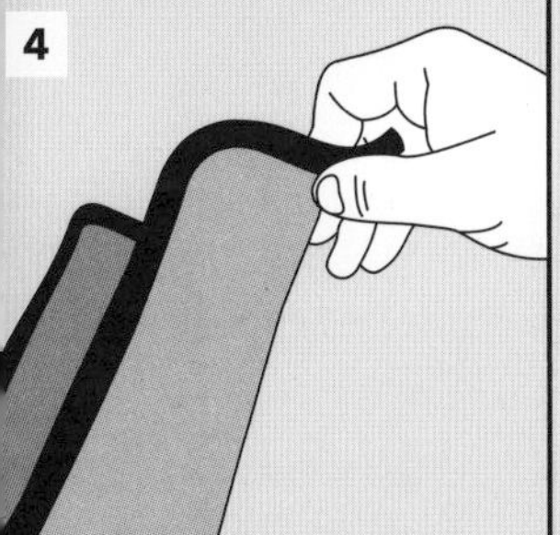

Use a single length of foam insulation to fold over the edges of the bottle until it is completely covered.

5

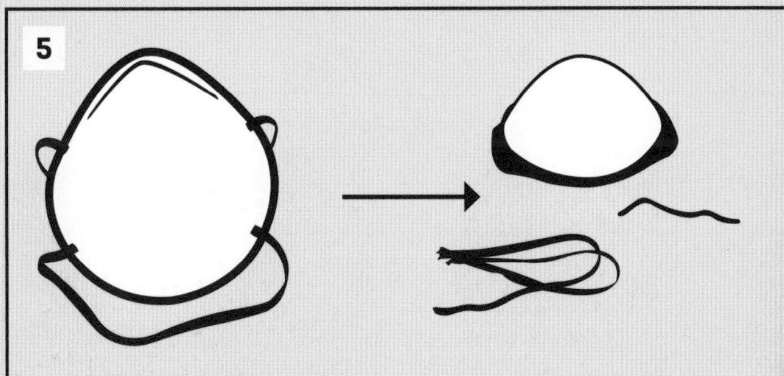

Remove the 2 elastic bands and metal bridge from the mask. Set the elastic bands aside and discard the metal bridge. Push the mask down into the neck of the bottle.

6

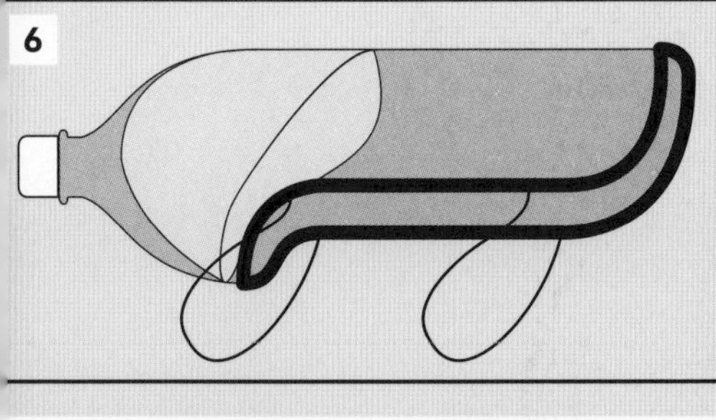

Make 4 small holes in the sides of the mask. Feed the ends of the elastic bands through the holes and tie them off so they can be pulled back through. Seal the holes with glue to prevent leaks in the mask. Carry a bottle of vinegar to soak the mouth cover before putting on the mask.

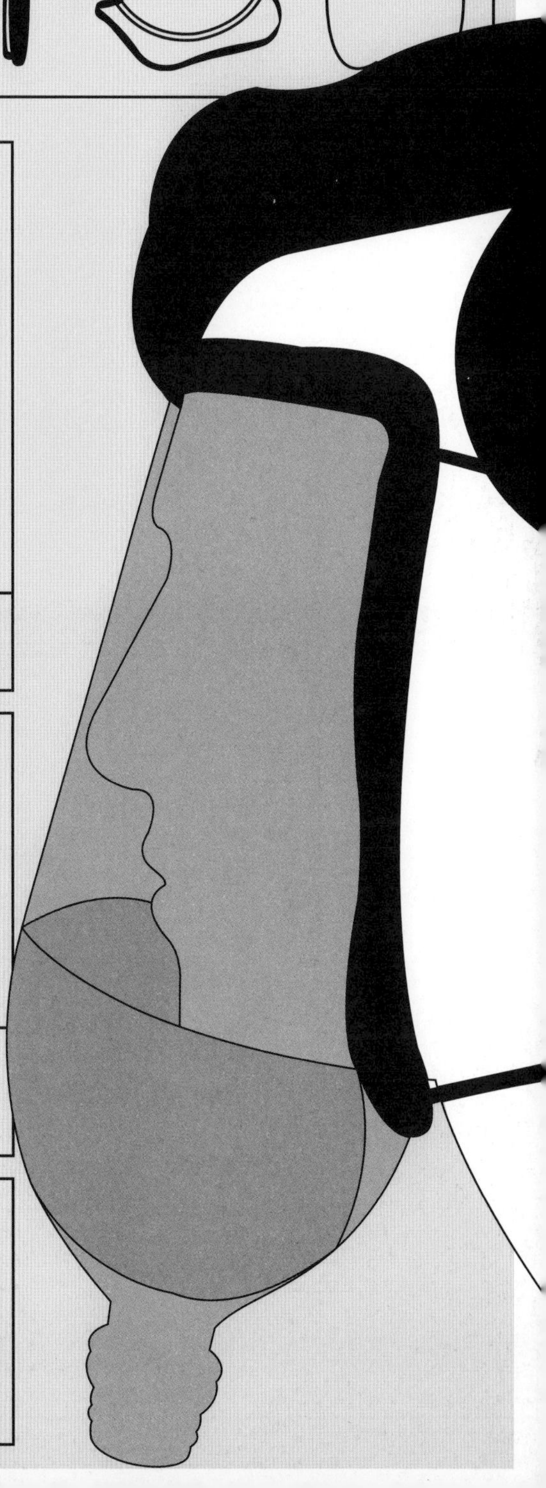

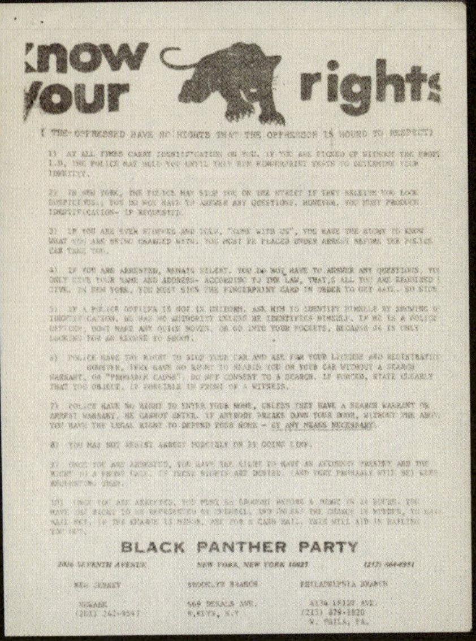
know your rights

(THE OPPRESSED HAVE NO RIGHTS THAT THE OPPRESSOR IS BOUND TO RESPECT)

1) AT ALL TIMES CARRY IDENTIFICATION ON YOU. IF YOU ARE PICKED UP WITHOUT THE PROPER I.D, THE POLICE MAY HOLD YOU UNTIL THEY RUN FINGERPRINT TESTS TO DETERMINE YOUR IDENTITY.

2) IN NEW YORK, THE POLICE MAY STOP YOU ON THE STREET IF THEY BELIEVE YOU LOOK SUSPICIOUS. YOU DO NOT HAVE TO ANSWER ANY QUESTIONS. HOWEVER, YOU MUST PRODUCE IDENTIFICATION- IF REQUESTED.

3) IF YOU ARE EVER STOPPED AND TOLD, "COME WITH US", YOU HAVE THE RIGHT TO KNOW WHAT YOU ARE BEING CHARGED WITH. YOU MUST BE PLACED UNDER ARREST BEFORE THE POLICE CAN TAKE YOU.

4) IF YOU ARE ARRESTED, REMAIN SILENT. YOU DO NOT HAVE TO ANSWER ANY QUESTIONS. YOU ONLY GIVE YOUR NAME AND ADDRESS- ACCORDING TO THE LAW, THAT,S ALL YOU ARE REQUIRED [illegible] GIVE. IN NEW YORK, YOU MUST SIGN THE FINGERPRINT CARD IN ORDER TO GET BAIL. SO SIGN

5) IF A POLICE OFFICER IS NOT IN UNIFORM, ASK HIM TO IDENTIFY HIMSELF BY SHOWING [illegible] IDENTIFICATION. HE HAS NO AUTHORITY UNLESS HE IDENTIFIES HIMSELF. IF HE IS A POLICE OFFICER, DON'T MAKE ANY QUICK MOVES, OR GO INTO YOUR POCKETS, BECAUSE HE IS ONLY LOOKING FOR AN EXCUSE TO SHOOT.

6) POLICE HAVE THE RIGHT TO STOP YOUR CAR AND ASK FOR YOUR LICENSE AND REGISTRATION. HOWEVER, THEY HAVE NO RIGHT TO SEARCH YOU OR YOUR CAR WITHOUT A SEARCH WARRANT, OR "PROBABLE CAUSE". DO NOT CONSENT TO A SEARCH. IF FORCED, STATE CLEARLY THAT YOU OBJECT, IF POSSIBLE IN FRONT OF A WITNESS.

7) POLICE HAVE NO RIGHT TO ENTER YOUR HOME, UNLESS THEY HAVE A SEARCH WARRANT OR ARREST WARRANT, HE CANNOT ENTER. IF ANYBODY BREAKS DOWN YOUR DOOR, WITHOUT THE ABOVE, YOU HAVE THE LEGAL RIGHT TO DEFEND YOUR HOME – BY ANY MEANS NECESSARY.

8) YOU MAY NOT RESIST ARREST FORCIBLY OR BY GOING LIMP.

9) ONCE YOU ARE ARRESTED, YOU HAVE THE RIGHT TO HAVE AN ATTORNEY PRESENT AND THE RIGHT TO A PHONE CALL. IF THESE RIGHTS ARE DENIED, (AND THEY PROBABLY WILL BE) KEEP REQUESTING THEM.

10) ONCE YOU ARE ARRESTED, YOU MUST BE BROUGHT BEFORE A JUDGE IN 24 HOURS. YOU HAVE THE RIGHT TO BE REPRESENTED BY COUNSEL, AND UNLESS THE CHARGE IS MURDER, TO HAVE BAIL SET. IF THE CHARGE IS MINOR, ASK FOR A CASH BAIL. THIS WILL AID IN BAILING YOU OUT.

BLACK PANTHER PARTY

[illegible] SEVENTH AVENUE NEW YORK, NEW YORK [illegible] (212) [illegible]

NEW JERSEY NEWARK (201) 242-9547

BROOKLYN BRANCH 569 DEKALB AVE. B,KLYN, N.Y

PHILADELPHIA BRANCH 4134 [illegible] AVE. (215) 879-1820 W. PHILA, PA.

1

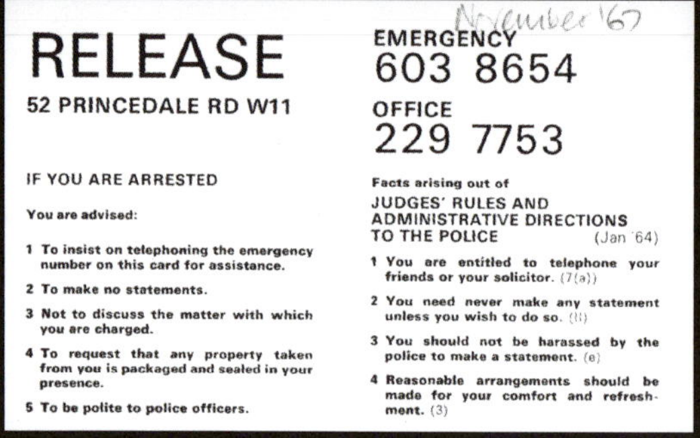
November '67

RELEASE

52 PRINCEDALE RD W11

EMERGENCY 603 8654

OFFICE 229 7753

IF YOU ARE ARRESTED

You are advised:

1 To insist on telephoning the emergency number on this card for assistance.

2 To make no statements.

3 Not to discuss the matter with which you are charged.

4 To request that any property taken from you is packaged and sealed in your presence.

5 To be polite to police officers.

Facts arising out of JUDGES' RULES AND ADMINISTRATIVE DIRECTIONS TO THE POLICE (Jan '64)

1 You are entitled to telephone your friends or your solicitor. (7(a))

2 You need never make any statement unless you wish to do so. (II)

3 You should not be harassed by the police to make a statement. (e)

4 Reasonable arrangements should be made for your comfort and refreshment. (3)

2

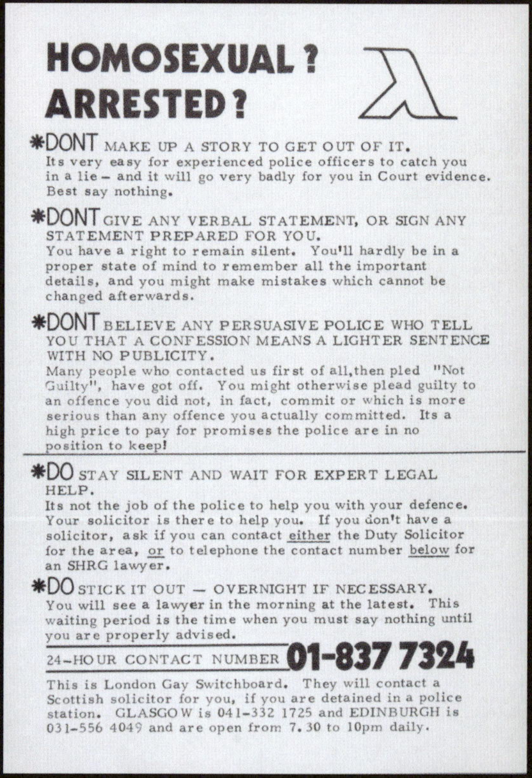
HOMOSEXUAL ? ARRESTED ? λ

✱DONT MAKE UP A STORY TO GET OUT OF IT.
Its very easy for experienced police officers to catch you in a lie – and it will go very badly for you in Court evidence. Best say nothing.

✱DONT GIVE ANY VERBAL STATEMENT, OR SIGN ANY STATEMENT PREPARED FOR YOU.
You have a right to remain silent. You'll hardly be in a proper state of mind to remember all the important details, and you might make mistakes which cannot be changed afterwards.

✱DONT BELIEVE ANY PERSUASIVE POLICE WHO TELL YOU THAT A CONFESSION MEANS A LIGHTER SENTENCE WITH NO PUBLICITY.
Many people who contacted us first of all,then pled "Not Guilty", have got off. You might otherwise plead guilty to an offence you did not, in fact, commit or which is more serious than any offence you actually committed. Its a high price to pay for promises the police are in no position to keep!

✱DO STAY SILENT AND WAIT FOR EXPERT LEGAL HELP.
Its not the job of the police to help you with your defence. Your solicitor is there to help you. If you don't have a solicitor, ask if you can contact either the Duty Solicitor for the area, or to telephone the contact number below for an SHRG lawyer.

✱DO STICK IT OUT — OVERNIGHT IF NECESSARY.
You will see a lawyer in the morning at the latest. This waiting period is the time when you must say nothing until you are properly advised.

24-HOUR CONTACT NUMBER **01-837 7324**

This is London Gay Switchboard. They will contact a Scottish solicitor for you, if you are detained in a police station. GLASGOW is 041-332 1725 and EDINBURGH is 031-556 4049 and are open from 7.30 to 10pm daily.

3

LEGAL ADVICE

STOP & SEARCH

Before a search you must be told the name and station of the searching officer, the reason you are being searched, and what power you are being searched under.

You do not have to give your name and address or answer questions under any stop and search power.

You are entitled to a receipt of the search.

GBC / LDMG Legal Observers are independent volunteers who gather evidence on behalf of anyone arrested and act to counter police intimidation and misbehaviour.

More info: www.ldmg.org.uk / www.greenandblackcross.org

gbclegal@riseup.net

4

6

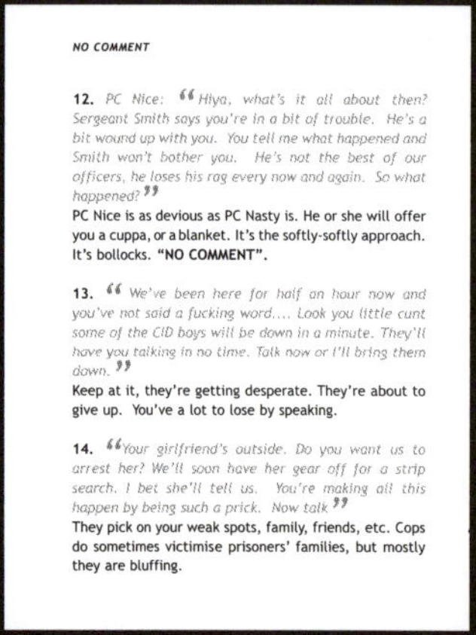
NO COMMENT

12. *PC Nice: "Hiya, what's it all about then? Sergeant Smith says you're in a bit of trouble. He's a bit wound up with you. You tell me what happened and Smith won't bother you. He's not the best of our officers, he loses his rag every now and again. So what happened?"*
PC Nice is as devious as PC Nasty is. He or she will offer you a cuppa, or a blanket. It's the softly-softly approach. It's bollocks. "NO COMMENT".

13. *"We've been here for half an hour now and you've not said a fucking word.... Look you little cunt some of the CID boys will be down in a minute. They'll have you talking in no time. Talk now or I'll bring them down."*
Keep at it, they're getting desperate. They're about to give up. You've a lot to lose by speaking.

14. *"Your girlfriend's outside. Do you want us to arrest her? We'll soon have her gear off for a strip search. I bet she'll tell us. You're making all this happen by being such a prick. Now talk"*
They pick on your weak spots, family, friends, etc. Cops do sometimes victimise prisoners' families, but mostly they are bluffing.

5

Above 1) Pamphlet, Black Panthers, New York, *c.*1968–9 • Black Panther Printed Ephemera Archive, Harvard University; 2) Bust card, Release, London, 1967 • Release Archive Warwick University; 3) Pamphlet, Scottish Homosexual Rights Group, Edinburgh, 1979 • SHRG Archive, National Records of Scotland; 4) Bust card, Green and Black Cross, London, 2012 • Private collection; 5) 'Defendant's Guide to Arrest, 3rd ed.', Legal Defence and Monitoring Group, London, 2006; 6) Billboard bust card, Space Hijackers, London, 2011 • 'Know your rights' pamphlets have a long history, but from the 1960s gay liberation and drug campaigning groups began producing small 'bust cards' to be carried in case of arrest. Release in London were first to use this term to describe them. The idea spread internationally. Following the increased criminalization of protest in some places, such cards have become increasingly common.

Opposite Occupy Sandy way-finder street sign, New York, 2012. When hurricane Sandy hit New York in October 2012, some Occupy activists turned their protest camp infrastructure towards disaster relief.

OCCUPY SANDY
RESTORING POWER TO THE PEOPLE
23
MIN WALK
INFORMATION
BASIC NEEDS
ELECTRICITY
VOLUNTEERS

flatpacked pop-up tents, photographic portraits, books, bodies, handles, gaffer tape, elastic ropes, foam rubber, cardboard shields, padding, inflatable life-rings, polypropylene ropes, empty water bottles, helmets, tyres, inflatable dinghies, videos, water pistols, book covers, overalls, Plexiglas shields, paint—**From *Tute Bianche* to the Book Bloc**

From 23 November 2010, first in Rome and London, then in Oakland, Berkley, Madrid, Montreal and elsewhere, rebellious students produced and organized what came to be called 'Book Blocs': groups in demonstrations with Plexiglas shields and helmets for bodily protection, pushing through police lines which enclosed a demonstration; resisting police dispersions of protests using baton charges; or pushing through police lines guarding the 'red zones'[1] of political power barricaded against the society outside. These practices endorsed the practices invented by the *Tute Bianche* (White Overalls) in Italy a little over ten years earlier. But much had changed since that time, from the expansion of the net economy and globalization in the 1990s to the bursting of the real-estate bubble and the onset of the world financial crisis in 2007. Europe is now afflicted by harsh austerity measures, while Italy in particular seems to be inexorably moving in the direction of economic, cultural and social decline. The Book Blocs were not merely an imitation of previous practices of conflict; rather, they reworked them with a new, powerful meaning. Body shields were replaced by shields fashioned as books, armour that made the point that knowledge could become a means of physical protection for those protesting against the dismantling of the public university, against job insecurity and unemployment. There are many more differences marking this new interpretation of the conflict. Let's continue in order.

Top Members of the *Tute Bianche* at the anti-International Monetary Fund and World Bank mobilization, Prague, September 2000.

TUTE BIANCHE: NEW PRACTICES OF CONFLICT AND COMMUNICATION

White overalls (typical work clothes) were used for the first time in Italy in September 1994 by the Ya Basta Association, during a demonstration staged by activists in Milan opposing the eviction of the historic Leoncavallo squatted social centre, which had opened in 1975. The Mayor had stated that 'Squatters are nothing other than ghosts now!' and the mass use of white overalls was an ironic return. But it was not until 1998 that the *Tute Bianche* became a political movement.[2] It took shape in Rome, looking to the protests of the unemployed in France. The choice of garment was a very precise one: compared to the blue overalls that are traditionally the garment of the working class, white overalls are the symbol of the youth workforce: mostly precarious, without rights or guarantees, excluded from the Fordist social contract with its permanent labour agreements, paid holidays and sick and maternity leave, and social security payments. A workforce with average qualifications, the result of the mass schooling that took place after 1968. These were the distinctive traits that defined the style of the *Tute Bianche*'s actions and agenda: blitzes with a high communication impact (the occupation of political and economic headquarters, eruptions into live television shows) that imposed visibility on what is invisible (job insecurity) and demanded social security benefits not related to job performance, the right to education and mobility. The *Tute Bianche* established a strong link between the practices of conflict and communication, identifying the mainstream media as the battleground in the belief that even radical conflict must define a positive tension and create consensus.

In 1999 the white overalls made their first appearance in street demonstrations. Along with the use of helmets and Plexiglas shields, white overalls became the symbol of a broader movement, which involved most of the Italian social centres. The aim of the *Tute Bianche* movement remained the same – to give visibility to the invisible – but the focus shifted to other issues: detention centres for migrants and the Kosovo War. In addition, the culture of such autonomous movements, which had developed in Italy's social centres, was influenced by American traditions of non-violent civil disobedience practices, from black civil rights movements to ecological blockades, and this mix was to produce new forms of protest and new ways of demonstrating.

Bodies and their performative power acquired a central role in the practices of struggle. The rallying cry of protest movements became 'bring your body into play': make it an instrument capable of violating neo-liberal legality, and at the same time make it the place where it is possible to collectively shape new ways of living. Padding was added to the Plexiglas shields and helmets. Sometimes it was of foam rubber, in other cases an assemblage of empty water bottles. Bodies were now protected and could confront the police in testudo formation, like the early Roman legions in times of siege. Sometimes the shields were replaced by inflatable dinghies, enormous inflatable life-rings or car or lorry tyres. Just like the shields, these objects were used to soften and muffle the blows inflicted with batons. In their hands protesters carried water pistols and rifles, which were used to distract the police by spraying coloured paint on to the visors of their helmets. Unarmed, inoffensive, brave, this 'army of dreamers', to use the words of the Zapatista movement – or 'beggars',[3] as the Luther Blissett (later Wu Ming) collective saw it – broke the silence, spoke up and created public space. Shields or helmets had been used before: the bike helmets of the German Black Bloc in the 1980s, the Japanese

'Those who do not move, do not notice their chains.' Rosa Luxemburg, attributed, 1910s

Zengakuren's wooden shields in the 1960s and, notably, the shields decorated with anti-apartheid graphics of New Zealand militants opposing the South African Rugby Union tour in 1981. But the *Tute Bianche*'s padding, both symbolic and practical, moved beyond a violent/non-violent strategic dichotomy. Confrontation with the police, who were so used to the ritual of more disordered clashes, now dramatized the movement's claims to justice and gave visibility and force to the issues often previously ignored by the media in its representation of such confrontation.

With the Seattle WTO protests of 30 November 1999 and the rise of alter- or anti-globalization movements, we witnessed another change. The civil disobedience practice of entering the red zone was aimed at the international summits of the world powers, from the Organization for Economic Co-operation and Development (OECD) and the World Bank to the International Monetary Fund and the G8, institutions that imposed economic policies on states, restricting welfare and social rights, and on businesses, causing downsizing and salary squeezes. The peak of this new phase was reached during the anti-G8 protests in Genoa in 2001.[4] The *Tute Bianche* movement decided to do without the symbol of its identity, the white overalls, but nevertheless attempted to violate the red zone. The repression in response was without precedent; the Berlusconi government, backed by the Bush administration, declared war on the alter-globalization movement: protester Carlo Giuliani was killed and hundreds of other protesters were tortured inside the Diaz school building and the Bolzaneto police station. The practice of civil disobedience suffered a significant setback, but in Italy the movement decided to roll out the conflict in other social areas (job insecurity, migration/citizenship, the common good), rather than challenge the summits of international powers: this was the beginning of the 'Disobedienti' (Disobedients) movement in Italy.[5] Elsewhere others were similarly inspired by these developments, from the WOMBLES in London, to Las Agencias's 'Prêt à Revolter' project in Barcelona, to Umsonst in Hamburg and Berlin.[6]

FROM THE *ONDA* MOVEMENT TO THE BOOK BLOC: THE ITALIAN STUDENT PROTEST MOVEMENT

The *Onda* (Wave, sometimes later the Anomalous Wave) student movement broke on to the scene in September 2008,[7] during the crucial phase of the world financial crisis. Following the failure of Lehman Brothers, and as the markets clamped down on credit, the international government response to this crisis was an exorbitant increase in public debt as resources were shifted to support the banks. The *Onda* responded, however, to the specific context of Italy's political debacle. In the spring of 2008 Silvio Berlusconi had again been elected Prime Minister. The government he formed through a pact with the Northern League, a xenophobic Italian political party, went to work, dealing with financial collapse by cutting welfare and slashing financial resources for public education, schools and universities, and for culture, art and entertainment. 'You can't eat culture' was the motto of the Minister for Economy, Giulio Tremonti, while the Minister for Welfare, Maurizio Sacconi, reminded new university graduates that they must get used to humble manual jobs. This direct attack on culture and intelligence distinguished the specific character of the Italian government's response to the systemic failure of capitalism.

It was in this grim setting that Italian students broke their silence, taking to the streets to fight back and boldly shout: 'We won't pay for their crisis!' Schools took the initiative: for the first time children, alongside their teachers and parents, played a leading role in overnight occupations. The fight was against the abolition of full-time schooling, the re-introduction of the single teacher per class and sanctimonious moralizing (inherent in the compulsory school smock uniform). Soon after, the lower schools also acted, and university students began their protest. Rome, in particular the city's Sapienza University, was the centre of the movement, which quickly fanned out to dozens of other universities across the country, spawning agitation and occupations. Prevailing over this more traditional form of protest – occupation – were wildcat demonstrations. Students poured out of their classrooms and campuses in oceanic waves, paralyzing the city by blocking the roads, invading railway stations and staging massive demonstrations. This generation with no future, condemned to uncertainty (in their work and emotional lives), ended the silence in 2008 and began to discover strength through a new form of action: the metropolitan strike, a strike by those without a workplace and its associated historic right to strike, blocking the city's movement, communication and production. The slogan chanted by the students was clear: 'If you block our future, we'll block the city!' On 30 October huge demonstrations began and continued for two months to undermine Berlusconi's popularity and the stability of his government.

Despite these mass protests with popular support, a budget bill was fast-tracked through parliament that summer. In 2010, as the bill was brought to parliament to be voted on, the student protest movement again broke out. This time, it was less isolated, quickly establishing links with working-class protests and other sectors of society being strangled by government policy, beginning with the migrant population. At the same time as the extraordinary wave of student protests in London, Italian students entered the Senate chamber for the first time in the history of the Italian republic and besieged the Chamber of Deputies and both houses of parliament for days while occupying the rooftops of faculties and monuments and blocking railways and streets. There was an unprecedented escalation in the conflict, which reached its peak on 14 December during a massive demonstration in Rome against the Berlusconi government. An estimated 100,000 protesters, students and temporary

'These kids don't have a little brother working in the coal mine, they don't have a little sister coughing her lungs out in the looms of the big mill towns of the Northeast. Why? Because we organized; we broke the back of the sweatshops in this country; we have child labour laws. Those were not benevolent gifts from enlightened management. They were fought for, they were bled for, they were died for by working people, by people like us. Kids ought to know that.' Utah Phillips, interview in *The Progessive*, 2003

workers besieged the government for several hours, clashing at times with the police. Anger over job uncertainty, unemployment and the growth of poverty erupted. From Rome to London the battle had commenced – for public education and against European austerity policies.

During this wave of protests, the universities became festive places of cultural experimentation. The defence of the public university entailed the invention of a new university: self-managed seminars, free universities, independent research laboratories – a proliferation of initiatives that attempted to transform from below the public university. During these exceptional days the Book Bloc made its first appearance.[8] The practice of holding up Plexiglas shields in demonstrations challenged the police's violence, but now to this was added a specific challenge to the education reform bill. Each shield was a book, a classic, a literary must: Petronius, Boccaccio, Deleuze, Spinoza, Morante, Miller, Machiavelli, the Italian Constitution and so on. The titles were chosen in the universities at the end of the assemblies or voted for online: books that had formerly been read and needed to be read once again, neglected classics of literature and philosophy that were no longer taught in programmes that reflected the impoverishment of the academy. The shields were made and painted in universities, in self-managed student houses and in social centres run mostly by students. Like the shields used by the *Tute Bianche*, book shields were made of Plexiglas, two sheets of which were used in the middle as reinforcement with cardboard and foam rubber padding; two elastic ropes served as handles. The fronts of the shields were emblazoned with the titles and spray-painted and varnished to resemble book covers. The website UniRiot.org (today UniCommon.org), a reference to the student movement, uploaded a video with instructions on making them that soon went viral.[9]

There were precedents, perhaps fortuitous, to these pictorial shields (but also in part inspired

by the *Tute Bianche*) in the photographic shields created by Las Agencias in Barcelona for the 'tactical embarrassment' of riot police during the 2001 demonstrations against the World Bank in Barcelona. On to those shields were laminated life-size photographic portraits of immigrant children, with stern faces and clenched fists. The police's orders to attack the demonstration required them to strike these images of unarmed children, and the psychological aspect of their violence against unarmed protesters was in this moment reflected back at them. Later, in London, at the 2007 Climate Camp against the expansion of Heathrow airport, similar photographs of the faces of those who had lost, or were about to lose, their homes as a result of climate change were attached to cardboard shields held at the front of a march to the headquarters of the British Aviation Authority. Once there, the shields were revealed to be boxes containing flatpacked pop-up tents, and a tent occupation in which the climate camp symbolically migrated to BAA's front entrance quickly began.

The Book Bloc's use of books as bodily protection to oppose the violence of the government and the police force by those who believed that knowledge is always an expression of freedom reflected Foucault's view that books serve to 'take position'.[10] With the book shields students won over the support of the wider public. Book Blocs proceeded to spread mimetically over the next year, surging in cycles of struggle across national borders

to make their appearance in London (7 December 2010),[11] Genoa (12 December), Milan (14 December), Umeå (15 May 2011), Oakland (18 June), Manchester (2 October), Berkeley (22 October), Madrid (17 November). They had a transversal, pluralistic appeal: everyone can choose his or her own book, everyone can make his or her own shield, everyone can recount his or her own personal rebellion.[12]

The *Onda* student protest had followed the path pioneered by the *Tute Bianche* movement – with special attention to the link between practices of conflict and communication – but spoke of something different. Not only did the *Onda* protest tactics vary, from occupation of rooftops and

Opposite Book Bloc, Rome, November 2010.

Top 'Artmani' photographic shields made by Las Agencias and used in protests against the World Bank, Barcelona, 2001. 'Artmani' translates as 'art for demos', but also puns on the 'Armani' brand. The faces depicted are those of Zapatista women, immigrant children and others.

Right Top Pictorial shields at Heathrow Climate Camp, August 2007.

Right Bottom Pictorial shields, as used at Heathrow Climate Camp • Photographic prints, cardboard, gaffer tape, polypropylene rope and pop-up tent • London, 2007 • Museum of London.

railway stations to faculty buildings and streets, but so too did their symbolic values. Whereas the *Tute Bianche* used shields as a symbol of the movement's

identity, the *Onda* protester could choose an individual identity. Students had combined with a youth workforce on temporary labour contracts to become leading players in social movements, pitted against two-dimensional politics.

READING THE CRISIS IN THE BOOK SHIELDS

What do these Book Blocs, appearing across international cities, have in common? Firstly, the context of the economic crisis that began in Britain and the United States in 2007, and then spread to the rest of the world, which was taken as an opportunity to launch a new and violent enclosure of the commons: the privatization of welfare and the salary humiliation of an entire generation.[13] Secondly, their supporters, the new poor: students, new graduates, short-term contract workers. Workforces newly qualified with knowledge and skills but no future and no rights, excluded from the social pact: 'I study hard, yet nothing lies ahead'; 'Despite many years of hard work I'm poorer than my parents.' Finally, the forms of protest they share, which unite a demand for democratic rights and a refusal of the hegemony of finance. The link between the claim for democracy and the redistribution of wealth is not accidental. We now live in a time when capitalism is radically separated from the expression of liberal democracy, and we only have to turn to China to grasp this. The uprisings that have flooded the global metropolises can do nothing other than speak in a new language, one that is both anti-capitalist and pro-democratic.

The Book Blocs represent what the technocrats of finance and the politicians call a 'lost generation': a feeling that goes beyond borders and national differences, initiating a new form of internationalism. This is a generation that will not surrender, that tries to tackle these radical transformations with an unprecedented synthesis of struggle

and knowledge, institutional innovation and reappropriation, conflict and communication. The book shields, in terms both symbolic and concrete, are the manifestations of the revolt: the material culture of movements that are against capitalist private property and look beyond the forms of public and state property. Just as knowledge and communication require social relations in order to exist, so the democracy claimed by the Book Bloc is free and common to all, an insurgent and pluralistic democracy in which bodies and books break boundaries and open themselves to life.[14]

Top Occupy Wall Street anniversary concert, Foley Square, New York, September 2012. Books maintained a symbolic importance in the Occupy movement, which followed shortly after these student movements. In the Occupy Wall Street camp in New York and in Gezi Park, Istanbul, 'people's libraries' were built and maintained.

Right Book Bloc, student and public-sector workers' protest against public funding cuts, London, December 2010.

Opposite Book Bloc, London, December 2010.

BOOK BLOC SHIELD

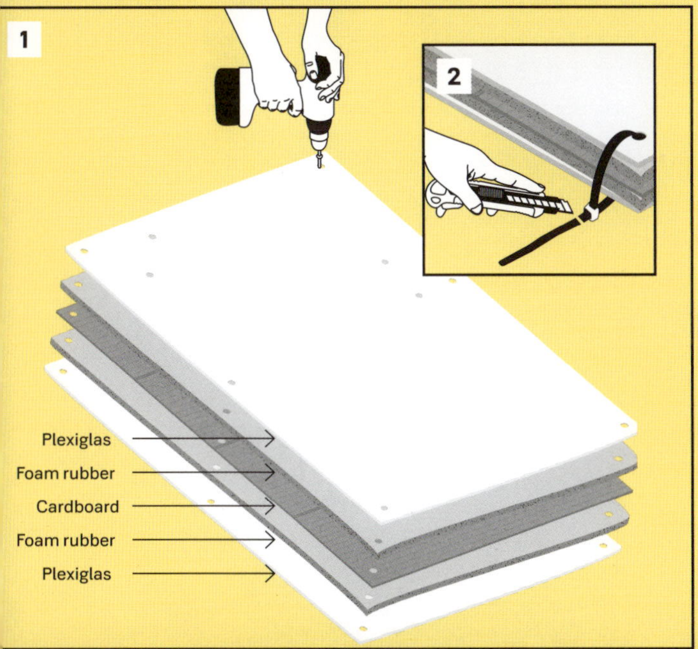

1. Layer the 5 sheets of material as shown above. Drill three holes on each long side of the block and two holes for the rope handles.
2. Secure the side holes with cable ties and cut off their ends.

Insert a single length of rope into the holes to form the arm handles.

Tie the ends of the rope from the front.

Cut off the ends of the rope from the front. Use paint or spray paint to draw the book cover artwork.

GEMEINSAM im KAMPF um
hafenstr. und
zusammenlegung
durchsetzen!

Di-rect Ac-tion

'There's a time when the operation of the machine becomes so odious, makes you so sick at heart, that you can't take part! ... And you've got to put your bodies upon the gears and upon the wheels ... And you've got to indicate to the people who run it ... that unless you're free, the machine will be prevented from working at all!' Mario Savio, Speech to the Berkeley Free Speech Movement, 2 December 1964

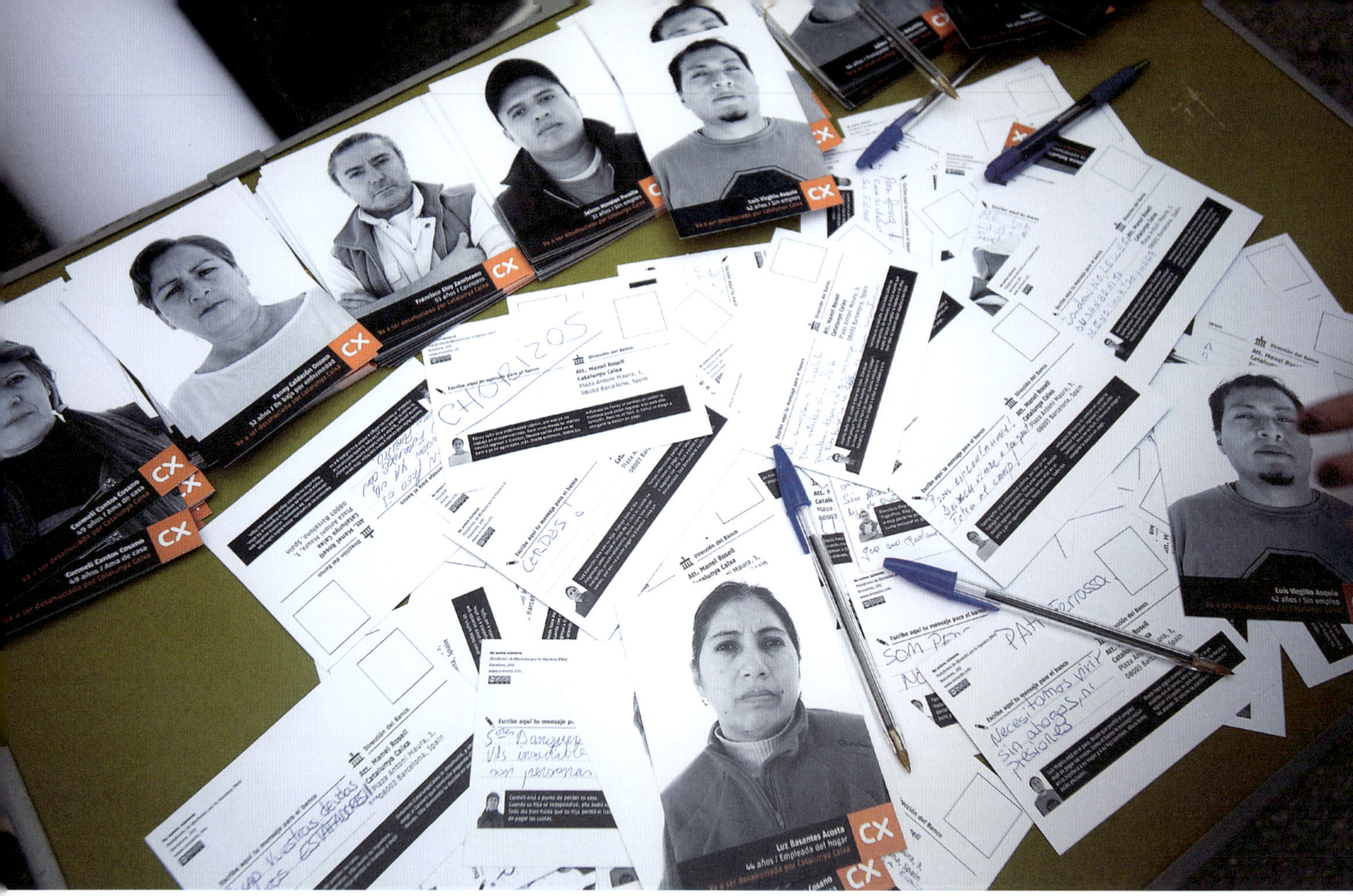

Previous Black Bloc protest, Hamburg, Germany, 1986. 'In response to these attacks [by police on squats in Hafenstrasse], the movement unleashed its own counteroffensive, marching more than 10,000 strong around a "black block" of at least 1,500 militants carrying a banner reading "Build Revolutionary Dual Power!" At the end of the march, the Black Bloc beat back the police in heavy fighting' (George Katsiaficas, *The Subversion of Politics*, Edinburgh, 2007, p.187). German media labelled the protesters 'die schwarze block' after their clothing. Different kinds of Black Blocs have appeared in movements internationally since, alongside pink, blue, book, childrens' and medieval blocs.

Top & Right TAF!/Enmedio, 'We Are Not Numbers' postcards, filled in during January 2013 protests in Barcelona organized by Platform for Mortgage Debt Victims against Caixa Catalunya, the bank that evicts most people in the region. People wrote personal messages such as: 'Thieves', 'You're taking our lives' and 'One day you will be judged'. They were pushed through the closed doors of the bank as images of people affected by mortgage debt were pasted on their walls.

Opposite Top 'Revolution of Dwarves' happening, Wroclaw, Poland, June 1988. The Orange Alternative formed under martial law in Communist Poland and carried out surreal happenings that played with the limits of the law (protests were illegal) and undermined the government's legitimacy. Ten thousand people attended this artistic happening, wearing orange dwarf hats and chanting 'We are the dwarves!' The militia was forced to follow instructions to round up and arrest all dwarves.

Opposite Bottom Flyer for 'Revolution of Dwarves' happening in Wroclaw, Łódź and Warsaw, Orange Alternative • Wroclaw • June 1988 • Orange Alternative Foundation.

KRASNALI

REWOLUCJA

Nasza Rewolucja ma zasięg ogólnopolski. W tym uroczystym dniu również stolica włączy się w nurt wydarzeń, posyłając w ramiona Rewolucji kwiat warszawskiej młodzieży.
Rewolucja Krasnoludków nie może odbyć się bez Ciebie! Jej los leży w Twoich rękach. Jeśli nie masz obu lewych- przygotuj się starannie. Duchową łączność ze Sprawą zamanifestujesz czerwoną częścią garderoby. Elementem krasnaliczznym może być również papierowa czapka w kolorze wiadomym. Pożądany wszelkiego rodzaju rewolucyjny ekwipunek w postaci gwizdków, grzechotek, trąbek. Mile widziane Sztandary Rewolucji jak również transparenty, wyrażające poparcie dla idei Rewolucji Krasnoludków jako takiej.
1.06. Płomień Rewolucji ogarnie całą Polskę. Tej szansy nie wolno nam zmarnować.

UWAGA

Osoby pozbawione czerwonych akcentów bądź dowodów poparcia dla Sprawy Rewolucji a przebywające tymczasowo w rejonie Wydarzeń narażone będą na szykany i represje ze strony STRAŻNIKÓW REWOLUCJI z racji wzbudzania niepokoju społecznego swą arewolucyjną postawą.

KOMITET OBRONY REWOLUCJI

Aparat fotograficzny jest również arewolucyjny. Zostaw go w domu i przyłącz się do nas.

K.O.R.

1.06. godz.16⁰⁰ Pasaż Śródmiejski

DAEWOO

LOCK-ON DEVICES

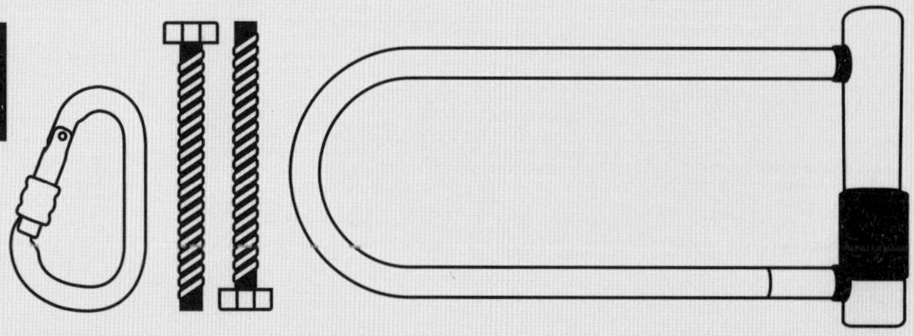

1

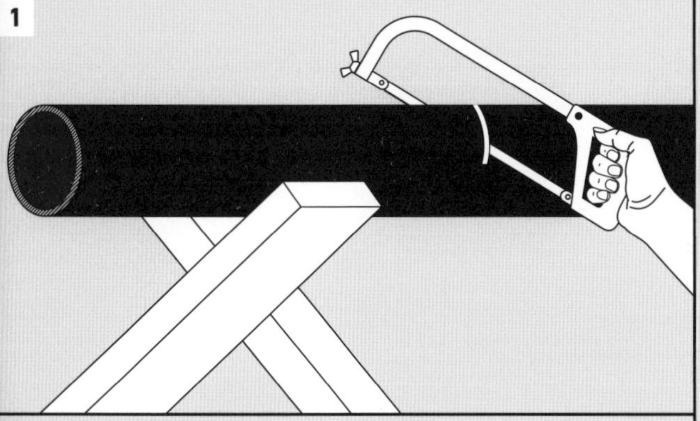

Use a hacksaw to cut the metal pipe into c.4-ft/1.2-m lengths – shorter ones are possible, but make access to the 'cuffs' easier.

2

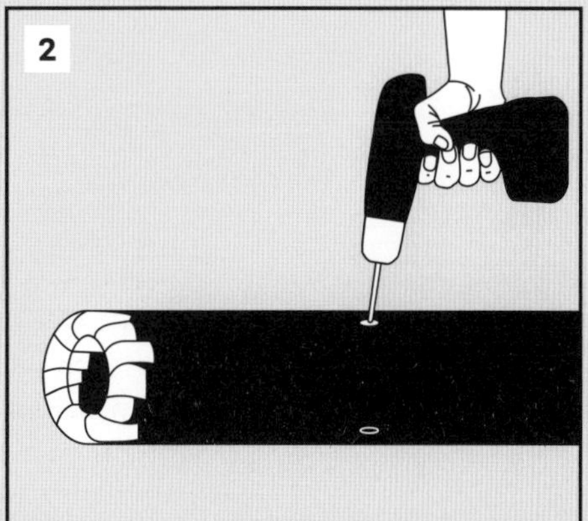

File down and pad the ends with foam and gaffer tape, and drill holes in both sides.

3

Put a bolt through the holes with nuts on the inside.

4

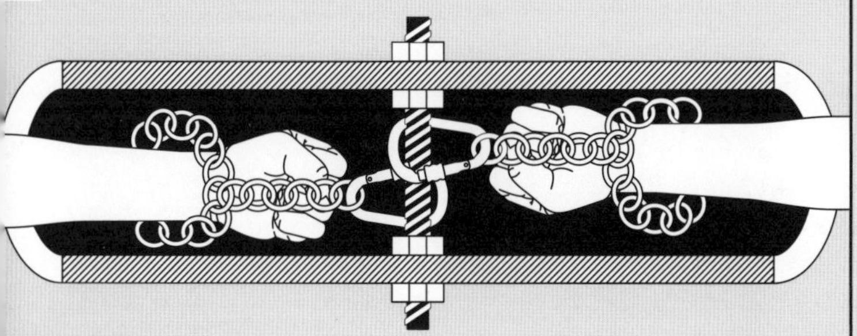

Wrap a chain around your wrist and attach a carabiner to the end of it. Insert your hand into the tube and lock the carabiner around the bolt.

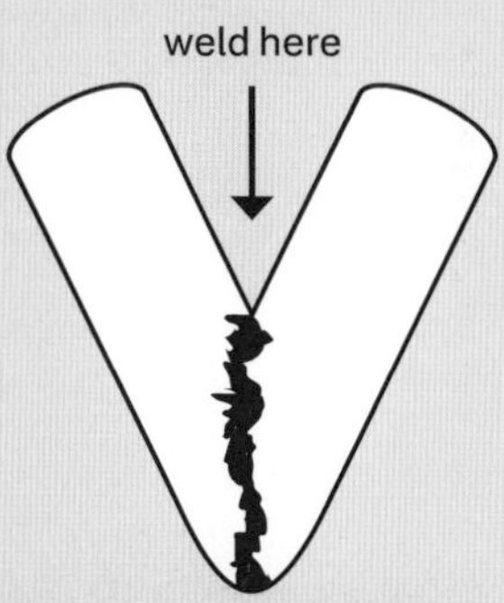

V-shaped tubes are an effective way for one person to secure him or herself around something.

General Safety & Support

- Dress for the weather. Stuff bin liners and pieces of roll mat down the back of your trousers.
- Plan a safe, comfortable lock-on position. Bring snacks.
- Have an affinity group support your safety and well-being, deal with media, security, etc.
- Plan ahead: how long will you lock-on for? Know the law. Practise media statements.
- Lock-on at the last practicable moment. Go to the toilet first.

BASIC LOCK-ON

Place a D-lock around the neck, then secure it to machinery to immobilize it or to gates to close them.

+ corrections

for use merely as a guide to where new edition has been corrected (on pp 2 + 3)

SCAFF TRIPODS

For your basic tripod, acquire:
3 scaff-poles, about 25 feet long;
2 swivelling scaff-clips;
and some rope — cheap, blue polyprop is fine. And a spanner for the nuts on the clips.

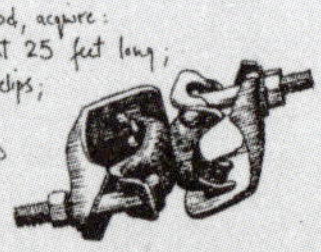

You also need a fairly large (high if indoors) space for fixing them, experimenting + practising.

It is tricky to get the clips fitted on so that the poles can lie parallel as well as erect into tripods.

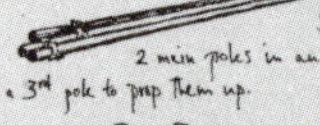

See the assembly as 2 main poles in an 'A' shape, plus a 3rd pole to prop them up.

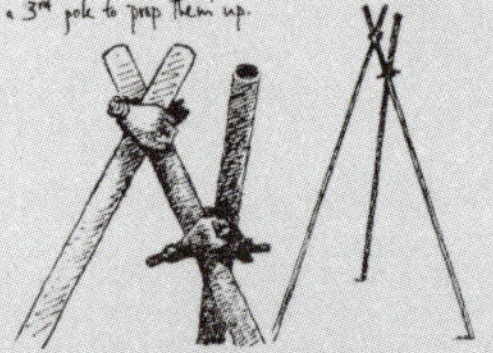

The clip for the 3rd pole has to be about a foot below the main poles' clip, to allow the main poles to close over it, and the central swivel of the clip should be at roughly 120° round the pole from that of the main clip. Experiment until you get the clips placed so that the poles can lie parallel + open into a tripod.

With iron poles you will probably need at least 5 people to erect it — which is done after first spreading the feet of the 'A', while simultaneously raising the 3rd foot:

At least one strong person to lift each of the 2 main legs by walking down beneath it from apex to base; one person to pull the 3rd leg out to prop up the 'A'; and one person with their foot braced against the base of each main pole to stop it skidding forward. These bracers must concentrate. If a foot slips the whole thing can crash down on someone's head.
With lighter aluminium poles you might manage with 3 people.

ONCE IT'S ERECTED, at least one person must shin up a pole at the speed of light, to be out of reach at the top.

A simple circumference rope tying the 3 poles together about 3 feet from the top can be fixed in advance and then stood on by up to 3 people.

A hammock sling is more comfortable and rather stylish. Make it out of a length of strong, light material, such as rip-stop nylon, knotted at either

end, with the 2 ends of a short rope tied securely just inside these knots. The rope can be slung over your shoulders as you shin up, and slipped over the poles at the top.

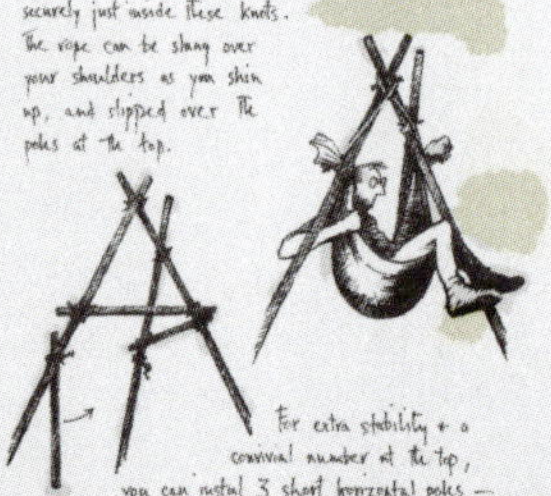

For extra stability + a convivial number at the top, you can install 3 short horizontal poles — with clips. Leave each short pole dangling from one clip until the tripod is up, then do up the 2nd clip. Don't forget the spanner. A climbing harness and slings make this job easier. The augmented tripod will be heavier. You may need more people to lift it.

A tripod lacking these extra bars can be stabilised against accidental slippage, though not against attack, with a circumference rope linking the legs 2 or 3 feet above the ground. Exhaust- or big jubilee-clips are useful to stop this rope riding up.

PRACTISE putting it up. Shambles develops into lightening expertise. When sorted out, paint 'L', 'R' + 'M' (left, right, middle) on the foot ends of the poles, the sides facing up as they lie ready to be raised. The 'M' pole should lie uppermost.
When the moment comes, rush the poles into the street and erect them fast fast fast behind a traffic-stopping banner.

Coat Loop Lock-ons

These haven't been used much as a tool in direct action campaigns, but they should be. They are very effective and also very cheap. They work by you wrapping your arms around something (an axle of a vehicle or a tree) and then fixing your wrists into loops fixed into your coat under your armpits. You then become pretty immovable but also, as in all direct action tactics, a bit vulnerable.

The material to use has to be strong and tough. We've used old seat belts and climbing tape. You'll need about a meter.

Sew the tape into your coat horizontally across the shoulder blades up to the armpits of your coat. Then double back the excess and sew the ends very firmly into place so they form loops. The bigger the loops, the easier they are to find in a panicky situation. The smaller they are the harder it is for them to pull your hands out (although you can twist the loops round and round so they tighten around your wrists).

It works as the tape goes all the way around your shoulder blades and so the pressure goes around your back rather than on the sewing under your arms. The loops are very difficult to get to, being under your garments and under your armpits.

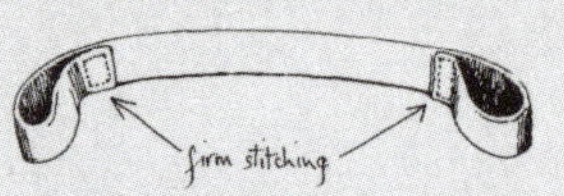

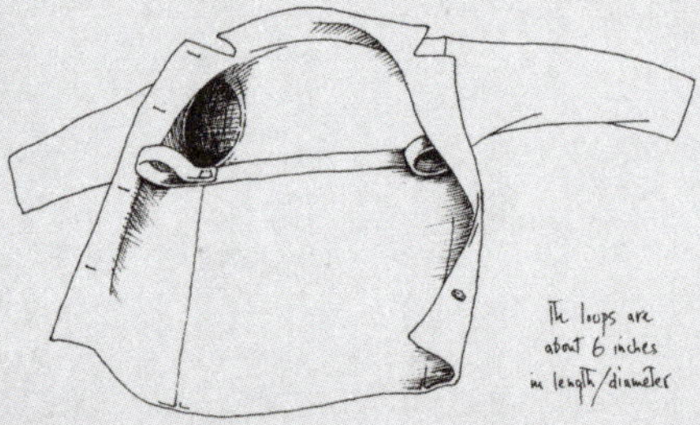

The loops are about 6 inches in length/diameter

Previous Top A Palestinian demonstrator, locked-on to a newly established Israeli separation barrier that annexed land in the West Bank village of Bil'in, March 2006.

Previous Bottom An activist locked-on during a 1995 protest in Stanworth Valley, Lancashire, against construction of the M65 motorway. The protest camp's eviction was the longest-running in post-war British history.

Opposite Working diagrams for scaffolding tripod designs and for coat-loop lock-on, drawn by B. Dahl • Pen, Tippex and glue on card • February 1997 • Private collection • The much-reproduced final versions of these diagrams appeared in *Road Raging: Top Tips for Wrecking Roadbuilding*, published by Road Alert! in 1997, during opposition to the building of the Newbury bypass in Berkshire, but the tripod designs were originally produced as a photocopied pamphlet.

Top Left First ever tripod blockade, New South Wales, Australia, 1989, used to blockade logging roads in order to oppose the renewal of a woodchipping licence to a private company. Some of these forests became National Parks.

Top Right Boom barrier at the June 1980 eviction of the Free Republic of Wendland, a camp of 1,000 protesters against a nuclear waste facility in Gorleben, Germany. The anti-nuclear movement has continued to grow globally. In 2002 an act legislated the closure of all German nuclear power plants by 2021.

Above Left A 100-ft/33-m scaffolding tower, Claremont Road, London, 1994, blocking construction of the M11 Link road. The tower was named Dolly, after one 93-year-old resident who refused to be evicted from her home in the street.

Above Right Banner suspended between tripods at the entrance to the 2009 Climate Camp, Blackheath, historic site of the 1381 peasants' revolt camp.

pneumatic drills, stilts, effigies, buckets of paint, paper, foam, foil, wire mesh, puppets, hats, black costumes, newspaper boxes, windows, dumpsters, hammers, teddy bears, poi, wooden and wire poles, balls, papier-mâché, backpacks, rags, silver cardboard, banners, inflatables, giant dresses, juggling clubs, face paint, sapling trees, bodies, carnival masks—**On the Phenomenology of Giant Puppets**

I begin with a simple observation. It's fair to say that if the average American knows just two things about the mass mobilizations of the anti-globalization movement, they are first, that they often involve people dressed in black who break windows, and second, that they involve colourful giant puppets.

I want to start by asking why these images in particular appear to have so struck the popular imagination. I also want to ask why it is that of the two, American police seem to hate the puppets more. As many activists have observed, the forces of order in the United States seem to have a profound aversion to giant puppets. Often police strategies aim to destroy or capture them before they can even appear on the streets. As a result, a major concern for those planning actions soon becomes how to hide the puppets so they will not be destroyed in pre-emptive attacks. What's more, for many individual officers at least, the objection to puppets appears to be not merely strategic, but personal, even visceral. Cops hate puppets. Activists are puzzled as to why.

If nothing else, these two observations mark a neat structural opposition. Anarchists in Black Bloc mean to render themselves anonymous and interchangeable, identifiable only by their political affinity, their willingness to engage in militant tactics and their solidarity with one another. Hence the uniform black costumes. The papier-mâché puppets used in actions are all unique and individual: they tend to be brightly painted, but otherwise they vary wildly in size, shape and conception. So on the one hand one has faceless, black anonymous figures, all roughly the same; on the other polychrome goddesses and birds and pigs and politicians. One is a mass, anonymous, destructive and deadly serious; the other is a multiplicity of spectacular displays of whimsical creativity.

If these paired images seem somehow powerful, I would suggest it is because their juxtaposition says something important about what direct action aims to achieve. Let me begin by considering property destruction. Such acts are anything but random. They tend to follow strict ethical guidelines: individual possessions are off-limits, for example, along with any commercial property that is the basis of its owner's immediate livelihood. Every possible precaution is to be taken to avoid harming actual human beings. The targets – often carefully researched in advance – are corporate facades, banks and mass retail outlets, government buildings or other symbols of state power.

Their property destruction is an attempt to 'break the spell', to divert and redefine. Consider here the words of the famous communiqué of the N30 Seattle Black Bloc (N30: November 30, 1999 – the abbreviation became a naming convention after J18 – the June 18, 1999 Carnival Against Capital in London), from the section entitled 'On the Violence of Property':[1]

> When we smash a window, we aim to destroy the thin veneer of legitimacy that surrounds private property rights. At the same time, we exorcise that set of violent and destructive social relationships which has been imbued in almost everything around us. By 'destroying' private property, we convert its limited exchange value into an expanded use value. A storefront window becomes a vent to let some fresh air into the oppressive atmosphere of a retail outlet (at least until the police decide to tear-gas a nearby road blockade). A newspaper box becomes a tool for creating such vents or a small blockade for the reclamation of public space or an object to improve one's vantage point by standing on it. A dumpster becomes an obstruction to a phalanx of rioting cops and a source of heat and light. A building facade becomes a message board to record brainstorm ideas for a better world. After N30, many people will never see a shop window or a hammer the same way again. The potential uses of an entire cityscape have increased a thousand-fold. The number of broken windows pales in comparison to the number of broken spells – spells cast by a corporate hegemony to lull us into forgetfulness of all the violence committed in the name of private property rights and of all the potential of a society without them.[2]

Property destruction is a matter of taking an urban landscape full of endless corporate facades and flashing imagery that seems immutable, permanent and monumental and demonstrating just how fragile it really is. It is a literal shattering of illusions.

What then of puppets? Again, they seem the perfect complement. Giant papier-mâché puppets are created by taking the most ephemeral of materials – ideas, paper, wire mesh – and transforming them into something very like a monument, even if they are simultaneously somewhat ridiculous. A giant puppet is a mockery of the idea of a monument:[3] its inapproachability, monochrome solemnity and, above all, its implication of permanence, the state's attempt (itself ultimately ridiculous) to turn its principle and history into eternal verities. If one is meant to shatter the existing 'spectacle', the other, it seems to me, is intended to suggest the permanent capacity to create new ones. It is the process of production that is really the point. There are brainstorming sessions to come up with themes, the wire frames lie on floors surrounded by buckets of paint and construction materials, almost never alone, with small teams moulding, painting, smoking, eating, playing music, arguing, wandering in and out. Everything is designed to be communal, egalitarian, expressive. The objects themselves are not expected to last. They are for the most part made of delicate materials; few would withstand a heavy rainstorm; some are even self-consciously destroyed or set ablaze during actions. In the absence of permanent storage, they quickly fall apart.

Their images constitute a universe. Puppetistas aim for a rough balance between positive and negative images. On the one hand, one might have the Giant Pig that represents the World Bank; on the other, a Giant Liberation Puppet whose arms can block an entire highway. Many of the most famous images identify marchers and the things they wear or carry: for instance, a giant bird puppet at the A16 (16–22 April 2000) International Monetary Fund/World Bank protests in Washington DC was accompanied by hundreds of little birds on top of signs distributed to all and sundry.

Top Performance at Bread and Puppet Theater, Glover, Vermont, United States, 1991. An Iraqi woman holds a dead body. Behind her is the threatening figure of a 'butcher' – archetypal characters used by the Bread and Puppet Theater to represent white faceless bureaucrats, paraded in multiple anti-war marches.

Opposite Street performance in protest against a nuclear power plant in Caorso, Italy, 1983, using props designed by Piero Gilardi, who was a key figure in Arte Povera (Poor Art). From the late 1960s he created work for political street theatre and demonstrations.

The most striking images are often negative: the corporate control puppet at the 2000 Democratic Convention in Los Angeles, operating Bush and Gore like marionettes; a giant riot policeman who shoots out pepper spray; and endless ridiculous effigies. The mocking and destruction of effigies is, of course, one of the oldest and most familiar gestures of political protest, but the positive images are afforded little more respect.

Here is an extract from my early reflections, jotted down after time spent at the Philadelphia Puppet Warehouse before the Republican Convention of the same year:

> The question I keep asking myself is: why are these things even called 'puppets'? Normally one thinks of 'puppets' as figures that move in response to the motions of some puppeteer. Most of these have few if any moving parts. These are more light moving statues, worn or carried. So in what sense are they 'puppets'?[4]

In fact, there's usually no clear line between puppets, costumes, banners and symbols, and simple props. Everything is designed to overlap and reinforce. Puppets tend to be surrounded by a much larger 'carnival bloc', replete with clowns, stilt-walkers, jugglers, fire-breathers, unicyclists, Radical Cheerleaders, costumed kick-lines or, often, entire marching bands.

Tony Blair's famous comment in 2001 that he was not about to be swayed by 'some travelling anarchist circus' was not taken, by many, as an insult.[5] The connection is significant; for now, the critical thing is that every action will normally have its circus fringe, a collection of flying squads that circulate through the large street blockades to lift spirits, perform street theatre, and also, critically, to try to defuse moments of tension or potential conflict. Here is a first-hand account:

> They [the puppeteers] joined a group that was blockading the building in which talks were being held. 'People had linked arms,' Zimmerman says. 'The police had beaten and pepper-sprayed them already, and they threatened that they were coming back in five minutes to attack them again.' But the protesters held their line, linking arms and crying, blinded by the pepper spray. Burger, Zimmerman and their friends came along – on stilts, with clowns, a 40-foot puppet, and a belly dancer. They went up and down the line, leading the protesters in song. When the security van returned, they'd back the giant puppet up into its way. Somehow, this motley circus diffused the situation. 'They couldn't bring themselves to attack this bunch of people who were now singing songs.'[6]

For all the circus trappings, those making and deploying giant puppets argue they are deeply serious. 'Puppets are not cute,' insists Peter Schumann, director of Bread and Puppet Theater, the group historically most responsible for popularizing the use of papier-mâché figures in political protest since the 1960s. 'Puppets are effigies and gods and meaningful creatures.'[7] Sometimes, they are literally so: as with the Mayan gods that came to greet delegates at the WTO summit meetings in Cancun in September 2003. Still, if giant puppets, generically, are gods, most are foolish, ridiculous gods. The process of producing and displaying puppets becomes a way to both seize the power to make gods, and to make fun of it.

The sacred here is the sheer power of creativity, the power to bring the imagination into reality. But it is also as if the democratization of the sacred can

only be accomplished through a kind of burlesque. The constant self-mockery is never meant to genuinely undercut the gravity of what's being asserted; instead it recognizes that gods, though human creations, are still gods, but that taking this fact too seriously might prove dangerous.

The N30 protests and actions in Seattle against the WTO were a turning point. They came as a surprise to most in the American government. The Seattle police were clearly unprepared for the sophisticated tactics adopted by the hundreds of affinity groups that surrounded the hotel and, at least for the first day, effectively shut down the meetings. The initial impulse of many commanders appears to have been to respect the non-violence of the actions.[8] It was only after 1pm on 30 November, after Madeleine Albright's call to the governor from her hotel demanding that he tell them to do whatever they had to do to break 'blockade 13', that police began a full-blown assault with tear gas, pepper spray and concussion grenades. Even then, many seemed to hesitate, while others, when they did enter the fray, descended into wild rampages, attacking and arresting scores of ordinary shoppers. In the end the governor was forced to call in the National Guard. While the media pitched in by representing police actions as a response to Black Bloc actions that actually began much later, having to bring in federal troops was an undeniable, spectacular, symbolic defeat. In the immediate aftermath, law enforcement officials – on a national and international level – seem to have begun a concerted effort to develop a new strategy. It seems their conclusion was that the police had not resorted to violence quickly or efficiently enough. The new strategy, however, appears to have been one of aggressive pre-emption. The problem was

Top A 20-ft/6.6-m articulated puppet of Chac, Mayan God of Rain, formed part of a demonstration by indigenous people protesting against the privatization of water during an anti-World Trade Organization mobilization, Cancun, Mexico, 2003.

Opposite Top Picasso's *Guernica* dances through the streets in pieces as part of a protest against the invasion of Iraq, New York, 2003.

Opposite Bottom Giant inflatable cobblestones, designed by Eclectic Electric Collective, were used during the General Strike, Barcelona, February 2012.

how to justify this against a movement that was overwhelmingly non-violent, engaged in actions that for the most part could not even be defined as criminal[9] and whose message appeared to have at least potentially strong public appeal.

If one looks at what happened during the months immediately following Seattle, the first thing one observes is a series of pre-emptive strikes, aimed at threats that never quite materialized. Here is one example among many: hours before the A15 (15 April 2000) protests in Washington DC against the IMF and the World Bank, police round up 600 marchers in a pre-emptive arrest and seize the protesters' convergence centre. Police chief Charles Ramsey loudly claims to have discovered a workshop for manufacturing Molotov cocktails and homemade pepper spray inside. Washington DC police later admit no such workshop existed (they'd actually found paint thinner used in art projects and peppers used for the manufacture of gazpacho). However, the convergence centre remains closed and many of the puppets inside are appropriated. From this moment, a key issue in the weeks before any mobilization became how to hide and protect the puppets.

The police had adopted a very self-conscious media strategy. Their spokesmen would pepper each daily press conference with wild accusations, well aware that the crime-desk reporters assigned

to cover them (who usually relied on good working relations with police for their livelihood) would normally reproduce anything they said uncritically, and rarely considered that it merited a story if afterwards the claims turned out to be false. This same period began to see increasingly outlandish accounts of what had happened at Seattle. During the WTO protests themselves, I must emphasize, no one, including the Seattle police, had claimed that anarchists had done anything more militant than break windows. That was the end of November 1999. In March 2000, three months later, a story in the *Boston Herald* reported that, in the weeks before an upcoming biotech conference, officers from Seattle had come to brief the local police on how to deal with 'Seattle tactics', such as attacking police with 'chunks of concrete, BB guns, wrist rockets and large capacity squirt guns loaded with bleach and urine'.[10] In June, *New York Times* reporter Nicole Christian claimed that Seattle demonstrators had 'hurled Molotov cocktails, rocks and excrement at delegates and police officers'. On this occasion, after picketing at their offices, the *Times* ran a retraction, admitting that according to Seattle authorities no objects had been thrown at human beings.[11] Nonetheless, the account appears to have become canonical. Each time there is a new mobilization, stories invariably surface in local newspapers with the same list of 'Seattle tactics' – a list that also appears to have become enshrined in training manuals distributed to street cops. Before the third Summit of the Americas in Miami in 2003, for example, circulars distributed to local businessmen and civic groups listed the following 'Seattle tactics' as actions they should expect to see on the streets once the anarchists arrived: 'wrist rockets, Molotov cocktails (many were thrown in Seattle) ... crow bars, squirt guns (filled with acid or

Top Medieval Bloc Teddy Bear Catapult, Deconstructionist Institute for Surreal Topology, Anti-FTAA mobilizations, Quebec City, April 2001 (confiscated by police).

urine)'.[12] Again, according to the local police force's own accounts, none of these weapons or tactics had been used in Seattle.

Predictably, by the time the first marches began, most of downtown Miami lay shuttered and abandoned. Here, too, puppets were singled out. In the months before the summit, the Miami city council actually attempted to pass a law making the display of puppets illegal on the grounds that they could be used to conceal bombs.[13] It failed, but the message was out. The Black Bloc in Miami actually ended up spending most of its time and energy protecting the puppets. Miami also provides a vivid example of the peculiar personal animus many police seem to have against large figures made of papier-mâché. According to one eyewitness report, after police routed protesters from Seaside Plaza, forcing them to abandon their puppets, officers spent the next half hour or so systematically attacking and destroying them by shooting, kicking and ripping the remains. One even put a giant puppet in his squad car with the head sticking out and then drove so as to smash it against every sign and street post available.

It's easy to see how one of the main concerns in the wake of Seattle would be to ensure the reliability of one's troops. As commanders discovered in Seattle, officers who are used to considering themselves guardians of public safety frequently baulk, or at least waver, when given orders to make a baton charge against a collection of non-violent 16-year-old white girls. These are, after all, the very sort of people they are ordinarily expected to protect. At least some of the imagery, then, appears to be designed specifically to appeal to the sensibility of ordinary street cops. This would help to explain the otherwise peculiar emphasis on bodily fluids: the water pistols full of urine, for example. This appears to be very much a police obsession. Certainly it has next to nothing to do with anarchist sensibilities. When I've asked activists where they think such stories come from, most confess themselves deeply puzzled. None has ever heard of anyone actually transporting human waste to an action in order to hurl or shoot it at police, or can suggest why anyone might want to. A brick, some point out, is unlikely to injure an officer in full riot gear, although it will certainly slow him down. But what would be the point of shooting urine at him? Yet images like this re-emerge almost every time police attempt to justify a pre-emptive strike. In press conferences, they have been known to actually produce jars of urine and bags of faeces that they claim to have discovered hidden in backpacks or at activist convergence sites.

It is hard to see these claims making sense except within the peculiar economy of personal honour, typical of any institution that, like the police, operates on an essentially military ethos. For police officers, the most legitimate justification for violence is an assault on one's personal dignity. To cover another person in effluent is obviously about as powerful an assault on one's personal dignity as is possible. We also seem to be dealing here with a self-conscious allusion to the famous image of 1960s protesters 'spitting on soldiers in uniform' when they returned from Vietnam – one whose mythic power continues to resonate to this day, despite the fact that there's little evidence that it ever happened.[14] It's almost as if someone decided to ratchet the image up a notch: 'If spitting on a uniform is such an insult, what would be even worse?'

'Human history began with an act of disobedience, and it is not unlikely that it will be terminated by an act of obedience.' Erich Fromm, *On Disobedience as a Psychological and Moral Problem*, 1963

Police are also regularly warned that puppets might be used to conceal bombs or weapons.[15] If questioned on their attitudes towards puppets, this is how they are likely to respond. However, it's hard to imagine that this alone could explain the level of personal vindictiveness witnessed in Miami and other actions – especially since the police who hacked puppets to pieces must have been aware that there was nothing hidden inside them. The antipathy seems to run far deeper. Many activists have speculated on the reasons:

David Corston-Knowles's opinion: You have to bear in mind these are people who are trained to be paranoid. They really do have to ask themselves

whether something so big and inscrutable might contain explosives, however absurd that might seem from a non-violent protester's perspective. Police view their jobs not just as law enforcement, but also as maintaining order. And they take that job very personally. Giant demonstrations and giant puppets aren't orderly. They are about creating something – a different society, a different way of looking at things – and creativity is fundamentally at odds with the status quo.

Daniel Lang's opinion: One theory is that the cops just don't like being upstaged by someone putting on a bigger show. After all, normally *they're* the spectacle: they've got the blue uniforms, they've got the helicopters and horses and rows of shiny motorcycles. So maybe they just resent it when someone steals the show by coming up with something even bigger and even more visually striking. They want to take out the competition.

Yvonne Liu's opinion: It's because they're so big. Cops don't like things that tower over them. That's why they like to be on horses. Plus, puppets are silly and round and misshapen. Notice how much cops always have to maintain straight lines? They stand in straight lines, they always try to make you stand in straight lines ... I think round, misshapen things somehow offend them.

Max Uhlenbeck's opinion: Obviously, they hate to be reminded that they're puppets themselves.

Let's return, then, to the notion of a 'puppet intervention'. In Philadelphia, on the evening of 1 August 2001, we organized a press conference at which one of the few puppetistas who had escaped arrest that morning was given centre stage. During the press conference and subsequent talks with the media, we all emphasized that the puppet crews were, effectively, our peacekeepers. One of their jobs was to intervene in, and defuse, situations of potential violence. If the police were really primarily concerned with maintaining public order, as they alleged, peacekeepers seemed a strange choice for a pre-emptive strike. But the *manner* in which puppets can be used to defuse situations

> 'Capital always dreams of a perpetual motion machine, work from energy without loss. But time is asymmetric: the future is not going to be like the past. Through our refusals, our insubordination, all the plans come to nothing, all the machines wear out, break down. Capital's contradiction is that the very agents that create the "fuck up" possess the energies it needs. Only *we* are in perpetual motion: eternally energetic, crafty, obedient, cowardly, insolent, revolting, but always in a motion that is the only source of work, development, surplus.'
> George Caffentzis, *The Work/Energy Crisis and the Apocalypse*, 1980

of potential violence is completely different from that employed by, say, protest marshals. Police tend to appreciate the presence of marshals, since marshals are organized into a chain of command that police tend to treat as an extension of their own – and which, as a result, often effectively becomes just that. Unlike marshals, puppets cannot be used to convey orders. Rather, they transform and redefine situations of potential conflict.

It might be helpful here to reflect on the nature of the violence police represent. A former Los Angeles Police Department officer writing about the Rodney King case, in which four LAPD police were charged with excessively assaulting an African-American construction worker in 1991, pointed out that 'Cops don't beat up burglars.' If you want to cause the police to be violent, the surest way is to challenge their right to define the situation. This is not something a burglar is likely to do.[16] This, of course, makes perfect sense if we remember that police are, essentially, bureaucrats with guns. Bureaucratic procedures are all about questions of definition. Or, to be more precise, the imposition of a narrow range of pre-established schema on to a social reality that is, usually, infinitely more complex: a crowd can be either orderly or disorderly; a citizen can be white, black, Hispanic, or an Asian/Pacific Islander; a petitioner is or is not in possession of a valid photo ID. Such simplistic rubrics can only be maintained in the absence of dialogue; hence, the quintessential form of bureaucratic violence is the wielding of the truncheon when somebody 'talks back'.

The details of this play of imagination against structural violence are endlessly complicated. For now I only want to emphasize two points. The first is that the line of riot police is precisely the point where structural violence turns into the real thing. Therefore, it functions as a kind of wall against imaginative identification. The second is that this juxtaposition of imagination and violence reflects a much larger conflict between two principles of political action. The first, a 'political ontology of violence', assumes that the ultimate reality is one of force. To be a 'realist' in international relations, for example, is about being willing to accept the realities of violence. The second could be described as a political ontology of the imagination. It's not so much a matter of giving 'power to the imagination' as recognizing that the imagination is the source of power in the first place. Anarchists level a systematic and continual challenge to the right of the authorities to define the situation. They do it by proposing endless alternative frameworks – or, more precisely, by insisting on the power to switch frameworks whenever they like. Puppets are the very embodiment of this power.

What this means in the streets is that activists are effectively trying to collapse the political, negotiating process into the structure of the action itself. To win the contest, as it were, by continually changing the definition of what is the field, what are the rules, what are the stakes – and to do so on the field itself. A situation that is sort of like non-violent warfare becomes a situation that is sort of like a theatrical performance or religious ritual, and might equally well slip back at any time.

Opposite Reclaim the Streets party, M41 motorway, London, June 1996. This is a very rare example of a puppet performance not just altering the mood of a protest, but also taking peaceful direct action. Under this stilt-walker's giant dress, camouflaged by loud sound systems, activists with pneumatic drills dug up the motorway and planted trees, which remained after the crowd dispersed.

journals, magazines, pamphlets, leaflets, ring-bound photocopies, sandpaper, glass, printed matter, ink, paper, covers, sealed envelopes, letterpress blocks, movable type, composing sticks, forme, rollers, linotype machines, typewriters, carbon paper, rulers, cameras, plates, mimeographs, photocopiers, computers, inkjet printers, proofs, screen-print stencils, catalogues, images, colour charts, acetates, squeegees, thread, glue, staples—
Unpopular Pamphlets

If print media has had an integral place in modern movements of art and politics, the journal and magazine are perhaps the pre-eminent instances. *La Révolution surréaliste*, *L'Internationale situationniste* and *Quaderni rossi*, to take three iconic radical periodicals, are something like the mobile ground upon which Surrealism, the Situationists and Italian 'workerism' came into being through time – the sites and means by which these currents and movements honed their ideas and aesthetic styles, established group coherence and gained purchase on the social imaginary. The point is aptly made by Guy Debord, and with an ennui that presents an amusing contrast to the hallowed tones that usually accompany talk of *L'Internationale situationniste*: 'Even the fact of publishing a slightly "regular" journal is very tiresome; and, at the same time, one of our only weapons to define and hold on to a base.'[1]

In plain terms, then, the journal and magazine are significant political objects. And yet, in this close correlation between movement and object, they reveal themselves to be just that little bit too obedient – ordered and contained by the requirements of a movement. Advocates for magazines will have numerous examples with which to challenge that assessment, but I make it as a useful means of contrast to the media form which is the topic of this essay: the pamphlet.[2]

'There is a "natural right" of resistance for oppressed and overpowered minorities to use extra legal means if the legal ones have proved to be inadequate. Law and order are always and everywhere the law and order which protect the established hierarchy; it is nonsensical to invoke the absolute authority of this law and this order against those who suffer from it.'
Herbert Marcuse, *Repressive Tolerance*, 1965

In the small press pamphlet, as we shall see, the correlation between object and movement is much less secure, allowing the object a more indeterminate, exploratory and critical character, both in relationship to its sites of publication and circulation, and the socio-material world more broadly conceived. Indeed, no longer subordinated to a movement, the pamphlet-as-object might become a 'comrade', to borrow Aleksandr Rodchenko's astonishing Constructivist formulation. As he writes in a letter home from the 1925 Paris International Exposition of Modern Industrial and Decorative Arts, 'The light from the East [the Soviet revolution] is not only the liberation of workers ... the light from the East is in the new relation to the person, to woman, to things. Our things in our hands must be equals, comrades, and not these black and mournful slaves, as they are here.'[3] Against the common image of Marxism as an ascetic order, here in Russian Constructivism are glimpsed moments of a communist material culture 'imbued with the deepest sense of Things', in Boris Arvatov's words; or what Marx calls 'the complete *emancipation* of all human senses and attributes' as we come to 'suffer' the object.[4] This is posited within and against the human estrangement from objects that Marx identifies with commodity fetishism, where '*all* the physical and intellectual senses have been replaced by the simple estrangement of *all* these senses – the sense of *having*'.[5] As a form of object, commodity fetishism, then, is also a form of identity or subject. To unpack that a little, Marx's analysis of commodity fetishism is not an account of how capitalism reduces humans to the status of objects, as it is almost invariably understood. Rather, his concept seeks to grasp how we become subjects, individuals estranged from the fully social and sensory potential of organic and inorganic life. In other words, the subject of commodity fetishism – to present it in capitalism's own elevated terms of freedom and individuality – is the 'liberty' of the '*restricted* individual, restricted to himself'.[6]

Returning to my particular case, such approaches to what I will call the 'communist object' encourage us to displace the linguistic content of the pamphlet and concentrate on its 'intensive expressiveness', as Arvatov has it, across the full range of its material forms and relations, where these are the literal materials of its physical composition, its technological affordances, visual design, sensory effects, and social and economic conditions of production and consumption. This material field is at once determining and emergent, orchestrated by capitalist social relations and, as such, shot through with contingency, conflict and politics.[7] It is amid these material relations that we can seek out a communist form that is adequate to

– perhaps even transcends – the critical orientations of a pamphlet's content, where textual content and material form come into highly various relations of resonance and conflict. That is a tall order for a short essay; here I will pursue only a limited number of these material features as they arise in consideration of the pamphlets of one small press, east London's Unpopular Books, an example of which features in the *Disobedient Objects* exhibition.

Unpopular Books was established in the late 1970s by Fabian Tompsett, in part a product of his involvement in the co-operative print shop scene and the Rising Free bookshop and press (which published the first single-volume English edition of Raoul Vaneigem's *The Revolution of Everyday Life*, a book that suffered from a poor knowledge of binding materials such that it became an 'autodestructive commodity', 'the perfect Situationist book: it fell apart as you read it').[8] Unpopular Books has published books and leaflets, but the pamphlet is Tompsett's preferred medium, a point he makes with reference to aspects of its textual and physical form, and its process of production: 'It's not bulky, ideally you can put it in your pocket easily. It's not going to take you too long to read, but it's long enough to get somewhere. And you can make it in all these different ways.'[9] This concise appreciation tips us into consideration of the pamphlet's material form and its specific manifestations in editions by Unpopular Books.

In contrast with the cumulative thematic concerns and sedimented intellectual habits of a journal, one of the defining experiences of reading a pamphlet is encountering a particular and focused discourse – not 'too long' but 'long enough' – that is unmoored from a familiar and pre-structured critical environment. It is a discursive fragment, an isolated unit that circulates without the institutional authority, infrastructure, and temporal pacing that order and distribute serial publications through time and across space. With Unpopular Books, these physical and formal features intersect with the broader social form of the 'public', which is articulated in the name of this press through the curious notion of the 'unpopular'. Both journal and formal political organization share the need to court and consolidate a sizeable public, in the mode of readership, market or membership; for the former, this requirement is largely determined by the financial demands of publication, while for the latter it is the dominant criteria for self-validation as a pertinent political entity. By contrast, the low production cost and the fragmentary, occasional form of the pamphlet frees it up from the journal's requirement of audience share. This makes it an ideal medium for a communist press that seeks, as does Unpopular Books, to challenge received political truths and the tendency of political groups and radical subjectivities toward self-flattery, while destabilizing any political community that the press itself may accrue from the prestige of critical prowess. The appeal of intellectual autonomy in this regard is readily appreciable, notwithstanding the common attraction to dogma in political circles, but such a wilfully unpopular approach to political community requires further elucidation.

Unpopular Books may seem a peculiar name for a communist publishing project, and yet one of its sources is a passage from Marx: 'Both of us scoff at being popular.' It is a remark made against the emerging cult of personality attending to Marx and Engels in the 1870s, and favours instead a formulation of communism as a distributed and self-critical process, a process that wards off any delimiting centre of attraction. The remark is part of an epigram to a text included in an early Unpopular Books edition: a ring-bound photocopy of David Brown's hitherto unpublished, typed text translation of Jacques Camatte's *Capital and Community*.[10] This is one of a dozen works published by Unpopular Books on communist theory that extend Marx's unpopular insight by forwarding communism as a critical movement immanent to the mutating limits of capitalist social relations, and not as a privileged political subject, organizational form, or repertoire of ideas.[11] The popularity of any of the latter serves to close down the mutating limits of communism to a delimited identity; or, as Unpopular Books has it (if I can generalize from a comment made against the enduring appeal of the Situationist International),

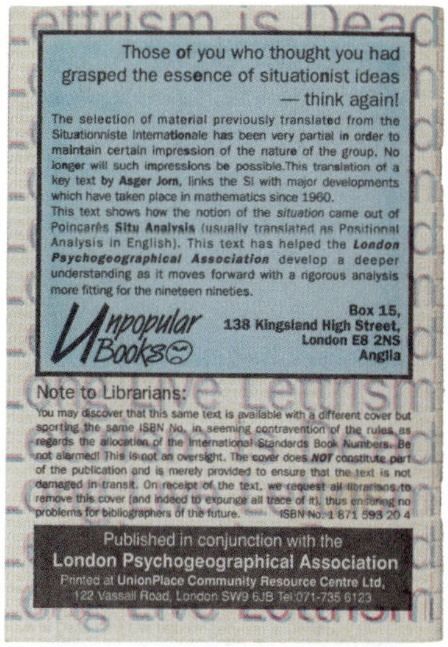

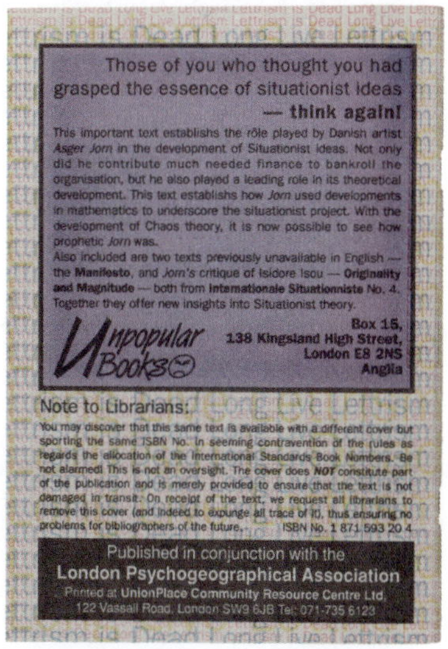

the 'popularity' of a particular critical form is an expression 'of the fact that the revolutionary movement has yet to overcome [its] weaknesses'.[12]

These capitalist social relations necessarily include objects, in critical relation to which lies another dimension of the unpopularity of the pamphlet. The press release for Guy Debord's *Mémoires*, a book of *détournement* famously bound in sandpaper, establishes the work as an 'unpopular book', doing so on the terrain of the commodity: 'in a time where all civilized nations battle to achieve the most popularity, using industrial design and mass-production of art objects and home appliances in the world market, a very unpopular book would be a much-needed rarity ... There is too much plastic, we prefer sandpaper'.[13] If this is not a direct source of the name of Unpopular Books, it is neither an unwarranted association, for Tompsett has played a significant role in the critical appropriation of Situationist thought, not least as (re-)founder in the early 1990s of the London Psychogeographical Association (LPA) (an organization originally established and folded simultaneously, when in 1957 it merged with others to found the Situationist International). Moreover, this line of reasoning against the 'popularity' of the book commodity had a place in the naming of Unpopular Books, playing as it did with the name and business model of the mass-market Popular Book Centre chain that was then common to the London high street. But if, like *Mémoires*, Unpopular Books editions constitute their communist form on the terrain of the commodity, they do so more in relation to the ideas and practices of Asger Jorn, who, contra Debord, maintained commitment to a communism of art and fabrication – to a 'materialist's love for matter', as Jorn put it.[14]

In the case of Unpopular Books, this appreciation of matter is manifest through engagement with the arts and conventions of printing and the means by which commercial and political value is articulated through printed matter. And so, in talking of the labour and value of printing, Tompsett comments: 'when you hear the term congealed labour you think of congealed ink. All the other printers do as well ... We would watch the printing press as the paper passed through it and imagine it squeezing value into these pieces of paper.'[15] The matter that is 'congealed' and 'squeezed' here is complex, having dimensions that are abstract as much as concrete, dimensions that can only be grasped with the aid of thought – with critique of the commodity form. That is, Tompsett's reference is to the concrete dimensions of abstract labour, where the circulation of the print commodity determines the form and value of the congealed labour and ink invested in its production. But

Top Asger Jorn, *Open Creation and its Enemies* (London: Unpopular Books, 1994) • Private collection.

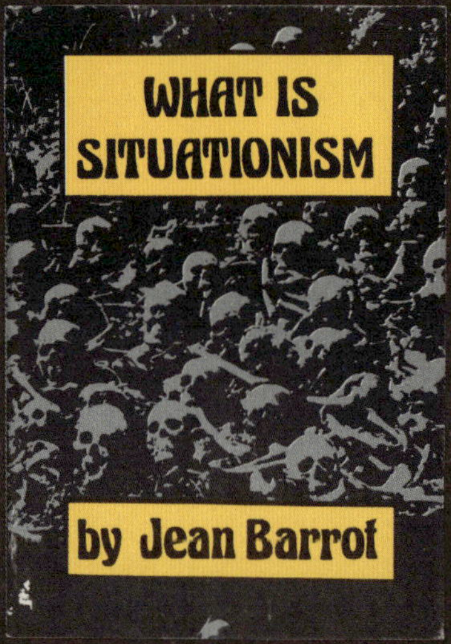
WHAT IS SITUATIONISM
by Jean Barrot

CLASS STRUGGLE in a GERMAN TOWN
Radicals intervene in a nuclear power plant construction in Phillipsburg

THE ECHO OF TIME
JACQUES CAMATTE

THE GREAT CONJUNCTION
A report by the London Psychogeographical Association and the Archaeogeodetic Association
The Symbols of a College, the Death of a King and the Maze on the Hill

GREEN APOCALYPSE
BY LUTHER BLISSETT AND STEWART HOME

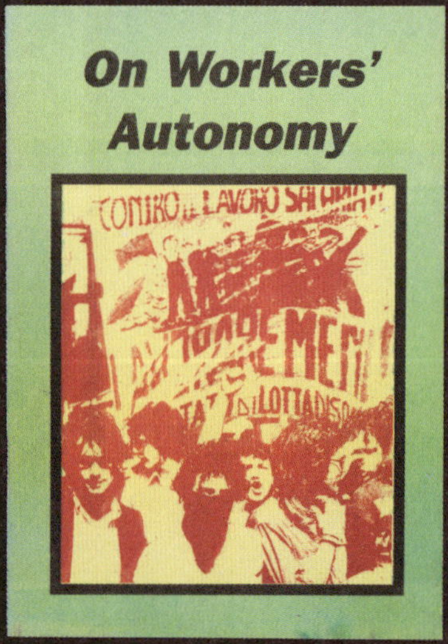
On Workers' Autonomy

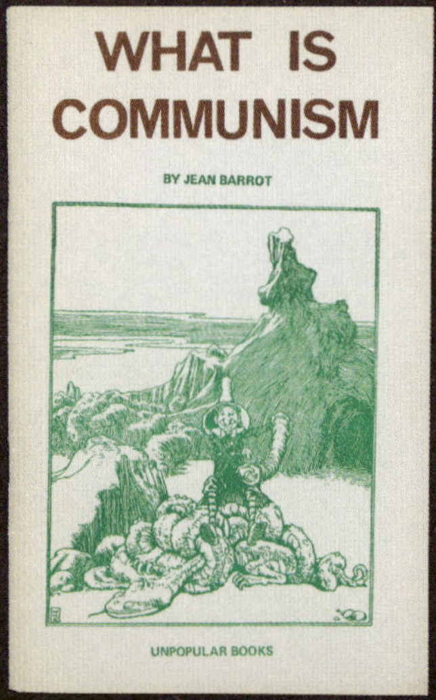
WHAT IS COMMUNISM
BY JEAN BARROT
UNPOPULAR BOOKS

The Revolution is Not a Masonic Affair
Boris Nicolaevsky's study of "Secret Societies in the First International"

MILITIAS:
Rooted in White Supremacy
A reprint of two articles from Turning the Tide, the journal of the California based People Against Racist Terror, published alongside Luther Blissett's essay 'Anarchism and the Militia' which exposes how neo-fascist apologists have used the anarchist movement to propogate their far-right agenda.

Tompsett is also gesturing toward something else in the printing process, a value of inks, papers and process that we might understand in terms of the intensive expressiveness of the communist object, what Jorn calls a non-monetary surplus value of matter in process.[16]

It is fitting, then, that the Unpopular Books pamphlet that is included in the *Disobedient Objects* exhibition, and which I take here as my example of this press's investment in matter and value, is a work authored by Jorn. *Open Creation and its Enemies* includes Tompsett's English translation of Jorn's texts 'Open Creation and its Enemies' and 'Originality and Magnitude (on the System of Isou)' with an introduction by Richard Essex (one of the numerous pseudonyms taken by Tompsett in his unpopular dissimulation of authorial property and prestige).[17] This pamphlet is constituted through reflexive attention to its material form in a fashion that repeats the bibliographic self-consciousness that Joad Raymond shows was a common motif of early modern pamphlet culture, with particular attention reserved for its commodity relations.[18] Unusually for a pamphlet, Tompsett had *Open Creation* allocated an International Standards Book Number (ISBN) and logged a copy with the British Library. In this manner it is placed and validated in the commercial field of the book as a uniform and determinate exchangeable commodity – and we should recall that it was precisely the standardizing properties of print technology that enabled the Gutenberg book to set the example of the modern commodity, being the first uniform and repeatable mass-produced object.[19] But *Open Creation* simultaneously troubles this regime, playing with the mechanisms that constitute it as a standardized and determinate entity. The pamphlet was printed in contravention of the ISBN allocation regulations with two different covers (though for consistency across the pamphlets – 'the particular mix of colours', 'the same moisture going into the paper' – they were set out simultaneously on the same A2 plates).[20] A 'Note to Librarians' on the back of each advises that the cover is merely a form of protection for the text in transit and should be expunged to avoid confusion for future bibliographers. And the inside covers each announce different Unpopular Books editions – *A Trip to Edzell Castle* and *An English Hacienda* – that remain unpublished, so introducing doubt into the authority and reliability traditionally associated with the act of print publication.

The *Open Creation* pamphlet is not, then, an autonomous entity wholly outside the structural patterns of the commercial book – such a formulation would not fit well with the perspective of the communist object, which, if it seeks to undo capitalist relations, is always operative within them. Instead, it achieves its particular intensive expression of process and value by operating as an unreliable mimic, opening a slight difference with the structures that constitute the book as commodity. At the same time, this pamphlet also cuts more directly against the commodity mode of the book. Unpopular Books' pamphlets have a price, but they have no exchange value; no capital was invested in order to realize surplus value from their sale. As to the labour involved in their fabrication, they have often been produced in the downtime between commercial print-runs. One could view this as a stolen moment of 'unalienated' work, but it is better understood as a contemporary instance of the strange unsettling of work and its identities that Jacques Rancière has characterized as the 'nights of labor'. This names the fleeting aesthetic activity pursued in the precious gaps between work by nineteenth-century worker-poets, -painters and -writers in their efforts to breach their separation from intellectual practice and so 'exorcise their

Opposite [left to right, top to bottom] Jean Barrot, *What is Situationism: Critique of the Situationist International* (London: Unpopular Books, 1987); Wildcat, *Class Struggle in a German Town* (London: AK Press and Unpopular Books, n.d.); Jacques Camatte, *The Echo of Time* (London: Unpopular Books, 1988); London Psychogeographical Association and Archaeogeodetic Association, *The Great Conjunction* (London: Unpopular Books, n.d.); Luther Blissett and Stewart Home, *Green Apocalypse* (London: Unpopular Books, n.d.); Pour un Intervention Communiste, *On Workers' Autonomy* (London: Unpopular Books, 1996); Jean Barrot, *What is Communism* (London: Unpopular Books, 1983); Boris Nicolaevsky, *The Revolution is not a Masonic Affair* (London: Unpopular Books, 1997); People Against Racist Terror and Luther Blissett, *Militias: Rooted in White Supremacy* (London: Unpopular Books, 1997) • Private collection.

inexorable future as useful workers' – a flight from the 'dictatorships ... of king work' that, paradoxically, reveals the impossibility of such a flight under the social conditions of capitalism.[21] Tompsett also frames the product and process of production of the pamphlet as constitutive of a 'potlatch' – the non-productive expenditure of the extravagant gift – though in contrast to the mutual obligation of potlatch economies, the LPA would surreptitiously place such pamphlets in book and charity shops encountered on their excursions, a mode of non-contractual distribution Tompsett calls 'negative shoplifting'.[22]

The 1994 *Open Creation* pamphlet is actually a revised edition; the preface notes a version published the year earlier in an issue of 50 to accompany a trip of the LPA to Calanais in the Outer Hebrides, an event associated with the commemoration of Jorn's death. There is a ritual dimension to the Calanais edition that introduces another facet to the communist form of the Unpopular Books pamphlet, a certain fetish quality. Unlike the revised edition, this pamphlet was not a mimic in the circuits of exchange of the book commodity, but an artefact exclusive to its event, where an event can be described, in the terms of *Open Creation*, as 'the constancy of intensity and the unique feeling of the propagation of the process'.[23] This event-driven existence is recorded in the pamphlet's preface with an account of its publication and distribution, and is consolidated with a copy held by Tompsett in a sealed envelope, posted from Calanais on the date of its publication.

We could go some way to theorizing these rare, ritual features with Deleuze and Guattari's concept of the 'monument', a precarious compound of matter and sensation that leaves behind everyday functionality and serves to affirm the event, carrying its reverberation into the future.[24] But Tompsett indicates a better route, having characterized such features of printed matter by using the category of the 'talisman', for its kindred form, the fetish, furnishes us with an anti-commodity fetishism appropriate for thinking the communist object.[25]

The fetish is an excessive object whose value is particular to an event, as contrasted to the value of the commodity, which lies in its universal exchangeability, emptied of all particular content. It is precisely in this contrast, at the intersection of heterogeneous value systems, that the term 'fetish' arose, in the efforts of seventeenth-century Dutch merchants to account for what they perceived to be the irrational attribution of value in West Africa to arbitrary objects. That which was valued was not the universally exchangeable object of money, but 'any "trifle" that "took" an African's "fancy"'.[26] But this is not an individualized value; the fetish, rather, is a destabilizing shock to the individual, to the subject. Recall that commodity fetishism is precisely the loss of the object, the alienation of human sensation in the isolated subject of private property. By contrast, the fetish is an object whose materiality 'threatens to *overpower* its subject', as Peter Pels has it, and this precisely because it lacks or interrupts everyday use and exchange values.[27] It is an object 'that has the quality to singularize itself and disrupt the circulation and commensurability of a system of values'.[28] And so, if the fetish is an object that confirms and prolongs an event, it is also an object that exists as event – not a discrete entity but an intensive materiality that emerges only in its disruptive intersection with commodity values. With this formulation, then, the second edition of *Open Creation* is drawn into the orbit opened by the first – both versions, now, being printed fetishes.

Denis Hollier has pursued aspects of this fetish quality – 'the irreplaceable, untransposable object' – of radical printed matter in his discussion of Georges Bataille's dissident Surrealist magazine, *Documents*.[29] In so doing, he draws attention to a trap that lurks in wait for the printed fetish: the reprint. For in being preserved and reiterated in the commercial publication circuits of Art History, the printed fetish loses its value: 'But it is for the kamikazes, for the most fleeting trackers of the avant-garde, those who have not even seen two winters, that the honor of the reprint is intended. He who wins loses.'[30] It is an astute observation that behoves me to consider the particular mode of preservation that befalls *Open Creation* by its inclusion in this exhibition. It has not been subject to the ignominy of the commercial reprint, but surely its form is violated in becoming an artefact of exhibition?

Allow me to approach this question at a tangent, by way of a parable drawn from a different communist publishing project, Infopool (established in 2000 by Jakob Jakobsen), in its encounter with Tate Modern. Availing themselves of the unpopular qualities of the pamphlet form, Infopool's members describe their pamphlets as 'self-institutional' entities – entities that are 'concerned with developing their own contexts'. It is a fledgling self-institution, infused with 'vulnerability', for without formal institutional structures or rights protection the pamphlets extend only a 'contract of "trust"' concerning sensitivity toward content and aim in an 'unprotected offer of communication'.[31] This may appear to be a weakness, but is in fact the pamphlet's singular power as a self-institutional object, since, in contrast to instances of political expression that are the products and bearers of institutional norms and regularities, this vulnerability affirms precisely the pamphlet's emergent quality, its existence only in the undetermined and exploratory 'institutions' that are articulated, or held, in its encounters.

Most likely, it was something of this self-institutional quality that appealed to the curators of the Tate's 2001 *Century City* exhibition, when they chose, without notification or consultation, to bind together three Infopool pamphlets in newly fortified covers, doctor the cover text and display the artefact threaded on a presentation wire. The museum's interest in these pamphlets exemplifies what Infopool call the 'valorisation of socialisation' – the commodification of social relations that seek to escape the commodity – that is common to contemporary cultural institutions, as they cast about for content and legitimacy.[32] In the Tate's case, it shows the inability of the institution to understand and handle the very qualities that had caught the curators' attention in the first place. For, in its new guise, the pamphlet's values of tentative and emergent self-institution were converted, with proprietorial disregard, into exhibition value; the value, as Arvatov has it, of 'murdered objects', 'hidden under glass'.[33] The only adequate response was for Infopool to liberate the artefact, documenting their 'Operation Reappropriation' with a damning critique of the museum's blunt and clumsy action:

> On display in a new hardback cover and threaded through with wire (the new vitrine) the pamphlets take on an aura that undermines both their form and content. They can no longer be passed on, given as gifts, and circulated to friends and fellow travelers i.e. to be self-institutional. In short the pamphlets have been commodified beyond their informal and nominal £1.00 price. The generator of value that is the Tate Modern has allotted them an immaterial cultural value (prestige, distinction) in exchange for the appearance of the value of their autonomy … We picked the pamphlets up on Friday February 9th. To negotiate their exit would have taken too long.[34]

There is disjunction and conflict, then, between the form of the pamphlet and the form of the museum, even before we evaluate the contemporary function of collection and exhibition in relation to the Victoria and Albert Museum's ever-present past as 'three-dimensional imperial archive'.[35] Yet, as I have argued, conflictual encounter is inherent to the form of the pamphlet as communist object – it has no uncomplicated, natural place. And the theme and content of *Disobedient Objects* demands a more critical and self-reflexive handling of the political object than that shown to Infopool by the Tate. Hence, as *Open Creation* rests in its vitrine, its communist form may well endure. Even if it does not, the pamphlet's failure will be in part an expression of its unpopular imperative as an object that invites its overcoming.

I WISH MY
BOYFRIEND WAS
AS DIRTY AS
YOUR POLICIES

Speak-ing Out

‘It was necessary to speak truth to power. Our concern is to reach all men, the great and the humble.’ American Friends Service Committee, *Speak Truth To Power: A Quaker Search for an Alternative to Violence*, 1955

Previous 'I wish my boyfriend was as dirty as your policies' placard, design by Coral Stoakes • Paint on cardboard • London, 2011 • V&A E.46–2014 • This placard was created for the 'March for an Alternative', a large demonstration against government spending cuts that was held in London on 26 March 2011. Students played a prominent role in the demonstration. The 'dirty policies' refer to cuts in education funding such as the Education Maintenance Allowance and the tripling of university tuition fees.

Top L.J. Roberts carrying their 'Gay Bashers Come and Get It' banner at the City Dyke March, New York, 2011. Some anti-gay protesters confronted the march, but were driven away as the banner elicited thunderous roars from the marchers.

Opposite 'We won't give it to Putin a third time' anti-government placard • Paint on cardboard • Moscow, 2012 • V&A E.47–2014 • The announcement in 2011 that Vladimir Putin was going to stand for a third term as president of Russia was met with mass anti-government demonstrations on the streets of Moscow. These were characterized by a panoply of handmade placards, which signalled a new creative turn in Russian protest towards wit and individual expression. This rainbow placard was made by a gay rights activist and plays on homophobic statements made by Putin.

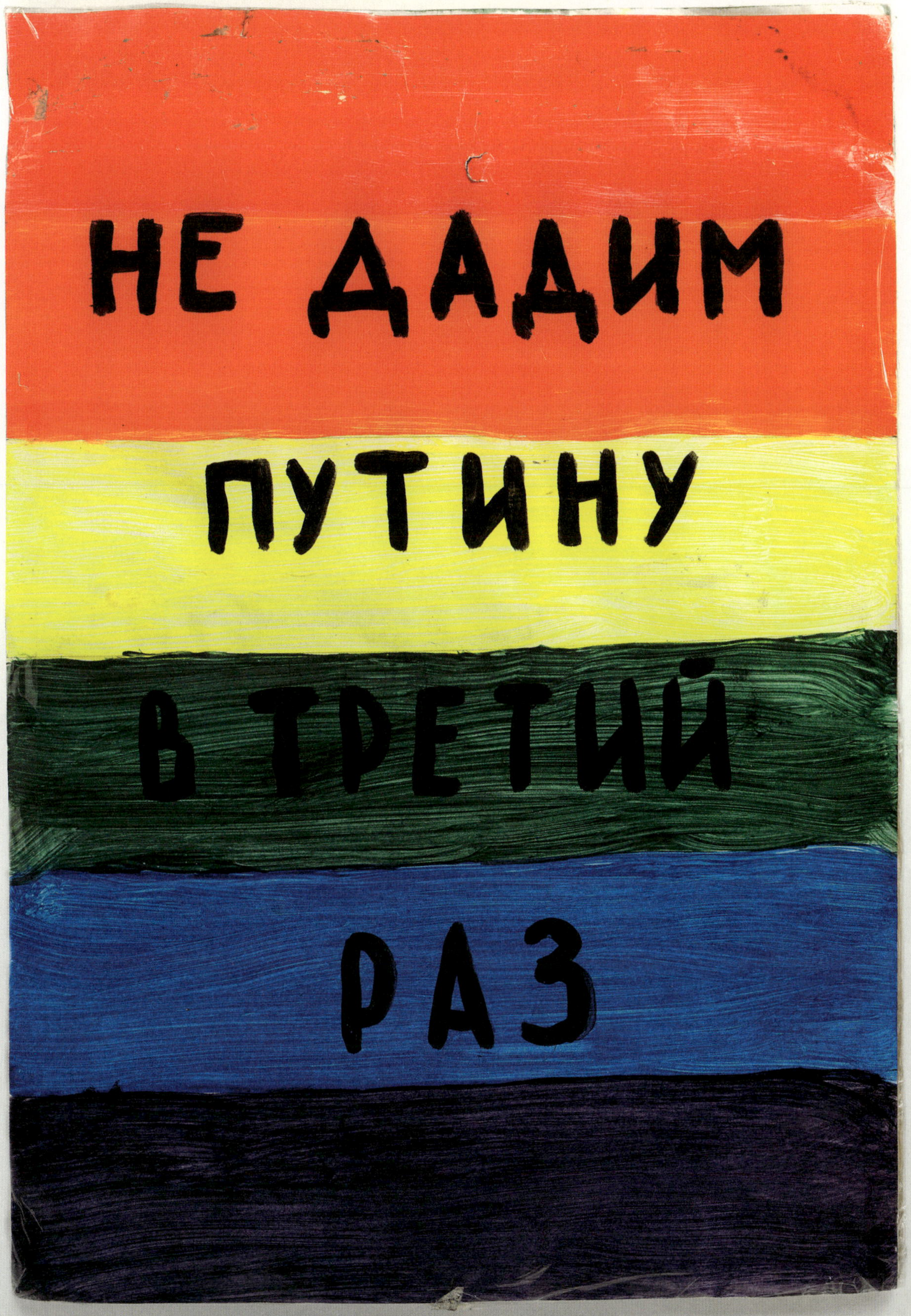
НЕ ДАДИМ
ПУТИНУ
В ТРЕТИЙ
РАЗ

Top Banner carried by Grand Legion of Incendiary and Tenacious Unicorn Revolutionaries (GLITUR) in Seattle, during 2012 Mayday Protests, which took place internationally in the context of the ongoing economic crisis.

Opposite Top Working drawing for 'Still the Enemy Within' banner marked up with comments from the South Yorkshire Community branch of Unite, Ed Hall • Pencil and pen on paper • Britain, 2013 • Private collection.

Opposite Bottom 'Still the Enemy Within' banner, designed by Ed Hall in 2013 for the South Yorkshire Community branch of Unite, shown marching through the street in Manchester, 29 September 2013, in a protest defending the National Health Service from cuts and privatization. Ed Hall has been designing and stitching marching banners for trade unions and campaign groups for over 30 years. His work is rooted in the traditional art of banner-making and his designs often focus on the dignity and history of the trade union represented. This banner references a number of famous past struggles, including the miners' strike and the protests against the poll tax, alongside present-day demonstrations against government spending cuts. In 1984, Margaret Thatcher had described striking miners and their families who resisted her reforms as 'the enemy within'.

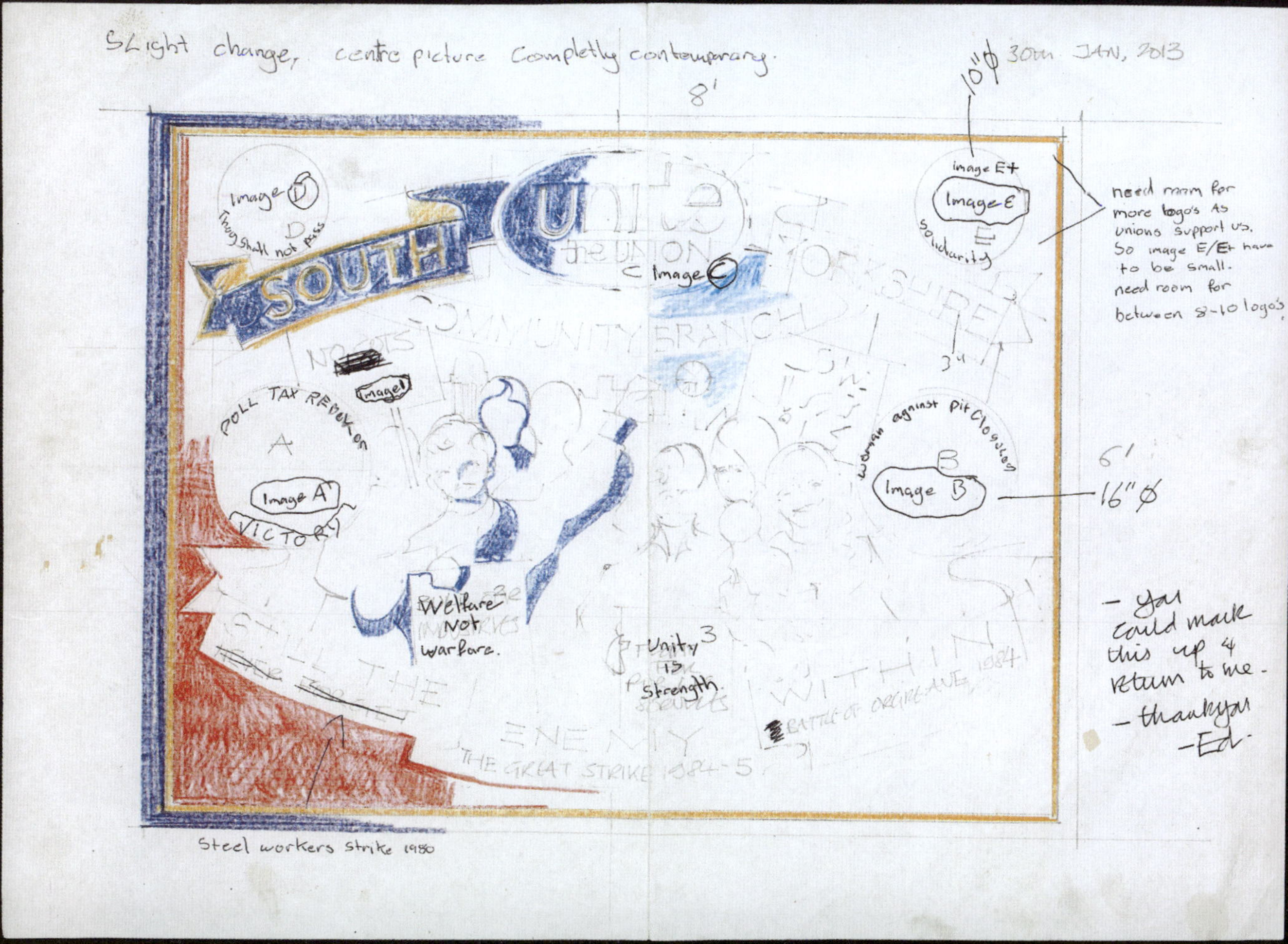
SLight change, centre picture Completly contemporary.
10"Ø 30TH JAN, 2013
8'
SOUTH
unite the UNION
YORKSHIRE
COMMUNITY BRANCH
Image D
They Shall not pass
Image C
Image E+
Image E
Solidarity
need room for more logo's As unions support us. So image E/E+ have to be small. need room for between 8-10 logo's.
3"
POLL TAX REBELLION
Image A
VICTORY
Woman against pit closures
Image B
6'
16"Ø
Welfare Not Warfare.
Unity is Strength.
STILL THE
ENEMY
WITHIN
THE GREAT STRIKE 1984-5
BATTLE OF ORGREAVE, 1984
- You could mark this up & return to me.
- thankyou
- Ed
Steel workers strike 1980

unite the UNION
SOUTH
YORKSHIRE
COMMUNITY BRANCH
WELFARE NOT WARFARE
UNITY IS STRENGTH
STILL THE
ENEMY
WITHIN
THE GREAT STRIKE 1984-5
BATTLE OF ORGREAVE 1984
dental

Right *Wapping Post*, produced by printers during the Wapping Dispute, 1986 • When print workers went on strike in January, Rupert Murdoch's News International fired 6,000 of them and moved production to a plant in Wapping that it had built in secret. Like the miners' strike, it was a turning point in British labour history in which the interests of big businesses sought to destroy the power of unions to represent workers' rights.

Bottom (left to right) *The Sun Right to Reply Special*, produced by Wapping printers in solidarity with striking miners, 1984 • *The Sun*, 1986 • *The Times Challenger*, published by the unions at *The Times* and *Sunday Times*, 1978 • During a 1978 dispute when this appeared, *The Times* itself was not published for a year. The dispute was the beginning of management attempts to reduce printers' pay and working conditions, partly effected through increased automation and the abandonment of traditional linotype printing methods.

Opposite Top *Evading Standards* newspapers, Reclaim the Streets, London, 1997, 1999 • The first *Evading Standards* newspaper was made for the March for Social Justice on 12 April 1997, a collaboration between Reclaim the Streets, Liverpool dockers and others. The papers were to be given out across London to publicize the march. However, undercover police spy DC Andrew Boyling had infiltrated the logistics group, and police seized and destroyed 25,000 copies and pre-emptively arrested distributors (the police seizures, property destruction and arrests were later deemed illegal in court). Despite this, a second design appeared a week later. A more widely circulated edition was printed to promote the 18 June 1999 Carnival Against Capital in London, whose style and organization was seminal for the 1999 Seattle WTO protests and subsequent global justice movement mobilizations.

Opposite Bottom Left *Class War*, Class War, 1985 • This cover represents a participant in the 1985 Handsworth riots, identifying black British unrest as also part of working class rebellion alongside the more celebrated cause of white working class miners.

Opposite Bottom Right *The Sun*, 'Murdoch Fucks Donkeys', anonymous, 1986 • Produced during the Wapping Dispute.

25p
WAPPING POST
WEEK-ENDING JUNE 7 VOL 2 NO 2
COP THAT
IT HAS cost the taxpayer almost £6 million to keep Rupert Murdoch's newspapers on the street.
Old Bill's £6 million truncheon voucher!
EXCLUSIVE by Murdoch NUJ strikers
HELLO MA'AM
The Queen gets her Wapping Post — see page 11
PLUS
The boycott spreads – page 2
What a mouth – Page 6-7
HELLO, HELLO
Meet the PC with an identity crisis – Page 5

NEWS OF THE WORLD
Sun
RIGHT of REPLY SPECIAL!
50p
Produced by workers on the Sun and News of the World
SOUVENIR EDITION!
In this issue
Eight page pull-out: NEWS FROM THE CHAPELS
JIMMY KNAPP
TONY BENN
PAUL FOOT ON THE MEDIA
All the news the papers won't print
WE WILL WIN!
Our cause is a noble one
By ARTHUR SCARGILL
£120,000 LIFELINE – Fleet Street cash for miners PAGE 16

THE Sun
Union Chiefs shock claim
I MAY BE GOD'S PROPHET
PLUS
IS GARY AN ARGY?
PLUS
Dirty Ken boots out Dirty Den
PLUS
A NEW SCUM RISING TODAY
Page Three Girl of the Year
You'll just love our gorgeous Ruperta!

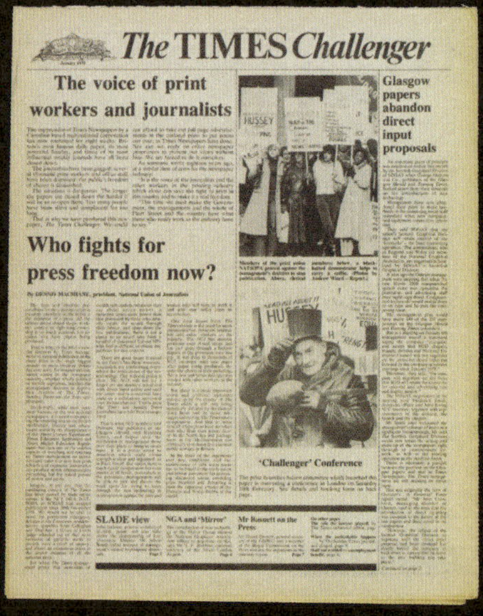
The TIMES Challenger
The voice of print workers and journalists
Glasgow papers abandon direct input proposals
Who fights for press freedom now?
'Challenger' Conference
SLADE view
NGA and 'Mirror'
Mr Bassett on the Press

SPECIAL EDITION

Evading Standards

LONDON, FRIDAY 18 JUNE, 1999 — INCORPORATING THE UNCENSORED NEWS — FREE

FREE trip to Cologne for G8 Summit celebrations. see page 19

WAR STARS Clinton and Albright's dark secret

NEW TODAY: INTERFERENCE FM – TUNE IN TO THE SOUND OF THE STREETS ON 107.4 FM

GLOBAL MARKET MELTDOWN

Panic stalks Square Mile following dramatic collapse of world financial markets.

By Watt Tyler and Emma Goldman

SHARES plummeted around the world today following last night's meltdown of Asian financial markets. Triggered by the collapse of Japan's national bank.

All through the night, investors worldwide have been fleeing equities and desperately seeking safety in gold and government bonds.

CITY WORKERS TOLD TAKE DAY OFF FOR CARNIVAL ON JUNE 18TH: FULL STORY PAGE 23

SPECIAL EDITION

Evading Standards

INCORPORATING THE UNCENSORED NEWS — FREE

Win a Dream Home! Do You Have The LUCKY NUMBERS — It's a lottery!

Never Mind the Ballots: vote SPICE!

THE EVENING SUN IS SETTING TOMORROW IS YOURS

GENERAL ELECTION CANCELLED

Election collapses as new polls reveal massive public cynicism

TV & RADIO BLACKOUT 68 — NO BUSINESS 36 — ENTERTAINMENT ENDS 17 — LIFE BEGINS 97

SPECIAL EDITION

Evading Standards

INCORPORATING NEW STANDARDS — FREE

Win a Dream Home!! Do You Have The LUCKY NUMBERS — It's a lottery!

Never Mind the Ballots: vote SPICE!

THE EVENING SUN IS SETTING TOMORROW IS YOURS

GENERAL ELECTION CANCELLED

Election collapses as new polls reveal massive public cynicism.

Dealers panic as collapsing election sends markets into chaos

TV & RADIO BLACKOUT 68 — NO BUSINESS 36 — ENTERTAINMENT ENDS 17 — LIFE BEGINS 97

THE Sun

THE PAPER THAT SUPPORTS OUR SCABS — SOUVENIR EDITION

WIN A SCAB IN YOUR SOARAWAY SUN

MURDOCH FUCKS DONKEYS!

Miss Emily Donkey

NEWSPAPER TYCOON IN BIZARRE SEX TRIANGLE

By KIERON SAUNDERS

YESTERDAY EVENING, in a late night raid on a printing complex in East London, the VICE SQUAD found Australian Press Baron Rupert 'Dingo-Dick' Murdoch, Metropolitan Police Commissioner 'Sir' Kenneth Newman and two donkeys practising bizzarre sexual relations.

We talked to the mother of one of the donkeys involved in this disgusting scandal, last night, at her home in Dollis Hill. Mrs Doris Donkey (54), said- "I am shocked that Emily should be involved with that revolting pervert Murdoch, I mean to say, its just not natural, is it?".

This incident gives a totally new meaning to the phrase:- 'The carrot and the stick.'

Mrs Newman and Son Kenny

We interviewed the mother of Police Commissioner, Mrs Pork Scratchings Newman. She said-" Darling Kenny told me it was all part of a Masonic Ritual and I see no reason to disbelieve him, and I'm sure his father would agree with me. That is, if anyone knew who he was!"

After extensive research, we failed to find any evidence of Rupert Murdoch having parents, but we have asked our correspondents down under to start looking under stones.

Rupert *Dingo Dick* Murdoch: Would You Trust This Man With Your Ass?

FORTRESS WAPPING

K1A

SAM BUSSES IN

By MURIEL BURDEN

Gushing, gorgeous Sam Fox, came down to our new plant in the cheeky East End of London yesterday to spread a little (not so little, eh!) joy and warmth to big Ruperts loyal workforce. Those pickets weren't half jealous, I can tell you matey!

Sam said:- " I'd love to get my hands on those new computers, but I'm as thick as two short planks."

What about those nasty pickets, we said?

"Whats a picket?, anyway, Ruperts paying me FIVE THOUSAND quid a week to scab, so sod that lot,"I'm alright Jack!"

Gosh, Sam, you're so smart, how do you do it? What is the secret of your success?

"Well, I just don't know, I think it must be my personality.....oh, yes, and I've got big tits!"

Cor, strewth, Sam, thats telling 'em!

★ SEE WHO'S WON FUCK ALL IN YOUR SUN BINGO ★

مصرف ليبيا المركزي
دينار واحد
صدرت بموجب القانون
عن مصرف ليبيا المركزي
المحافظ
1
7 ج/4 654924

TG4506835
မြန်မာနိုင်ငံတော်ဗဟိုဘဏ်
တစ်ကျပ်

THE U.S. DISPARITY of WEALTH
FEDERAL RESERVE NOTE
THE UNITED STATES OF AMERICA
THIS NOTE IS LEGAL TENDER FOR ALL DEBTS, PUBLIC AND PRIVATE
RICHEST
400
AMERICANS
F 82647456 A
BOTTOM
150,000,000
AMERICANS
WASHINGTON, D.C.
SERIES 2009
ONE DOLLAR
OCCUPYGEORGE.COM

There is a long history of defacing coins and banknotes in order to slip a political message into circulation. Sometimes this is done as a direct and immediate gesture of graffiti. In other cases there is a more involved design process. A currency is a highly controlled symbol of a state and tampering with it is a small but powerful act of subversion.

Opposite Top Libyan banknote defaced with marker pen • Printed paper • Libya, 2011 • Private collection • When Muammar Gaddafi fell from power in 2011, the Libyan people were left with his image on their banknotes. In many cases his face has been scribbled out – a miniature act of defiance that echoes the tearing down of statues of the dictator.

Opposite Middle Burmese one-kyat 'democracy note' • Printed and watermarked paper • Burma, 1989–90 • Private collection • When Aung San Suu Kyi won an overwhelming election victory in Burma in 1990, the military junta ignored the result and placed her under house arrest. Displaying her image, even in private, became grounds for arrest. At this time a designer working on a new banknote featuring General Aung San – the father of Aung San Suu Kyi – decided to subvert the commission. In the portrait drawn for the watermark, the features of the general have been subtly softened to resemble the face of his daughter. For a few months before the government realized and withdrew the 'democracy note', people in Burma could hold up their banknotes to the light and see a hidden portrait of the opposition leader.

Opposite Bottom 'Richest 400, Bottom 150,000,000', stamped US one-dollar banknote, design by Occupy George • Printed paper, hand-stamped in red ink • United States, 2011 • V&A: E.53–2014 • Inspired by the Occupy movement that began in New York City in 2011, artists Ivan Cash and Andy Dao (Occupy George) created a set of fact-based stamps illustrating wealth disparity in America. They could be found stamping the messages on to dollar bills throughout protests in San Francisco, Oakland and Berkeley. The designs were released on the Internet enabling anyone to participate. The very money that has divided America was utilized to communicate the heart of the problem. In this example it is revealed that the richest 400 people in America have as much wealth as the poorest 150,000,000.

Top British and Irish coins, stamped with paramilitary slogans • Metal • Northern Ireland, 1970s–90s • Linen Hall Library Collection • Stamping coins with paramilitary slogans was prevalent in Northern Ireland from the 1970s and throughout the period of the Troubles, especially in border counties where both British and Irish currencies were used alongside each other. The coins were generally stamped by metal workers and mechanics in their tea breaks as a popular everyday practice of political expression rather than an organized campaign. Two of the Irish coins bear the acronyms of the Loyalist paramilitary group the Ulster Volunteer Force and the Ulster Defence Association. A later Republican example from the 1990s has RIRA (Real Irish Republican Army) stamped over the head of the Queen on a British pound coin.

Top 'Dazibao' (large character) poster, Seoul, 1987. The poster announced the death of student protester Lee Han-Yeol. On 9 June 1987 Lee Han-Yeol had participated in demonstrations against the dictatorship in South Korea. A tear-gas canister fired by riot police penetrated his skull, eventually killing him. The public outcry was an essential part of the June Democracy Movement that brought down the regime. Lee Han-Yeol's funeral became a political event attended by over a million people.

Opposite Top 'Regime Change Begins at Home', set of cards, design by Noel Douglas/Bookmarks Publications • Print on waxed card • London, 2004 • V&A: E.52-2014 • These cards respond to a 2003 set issued to soldiers in Iraq by the US military listing the names, roles and addresses of members of Saddam Hussein's government, which had itself become a pro-war collector's item.

Opposite Bottom Graffiti produced by GraffitiWriter, United States, *c.*1999. GraffitiWriter is a 'robotic objector', developed by the Institute for Applied Autonomy (IAA) in 1999. A remote-controlled vehicle with a rack of spray cans mounted on the rear, it writes text on the ground in a manner similar to that of a dot-matrix printer. It allows the operator to maintain a safe distance from the act of graffiti while the seductive power of technology confers a sense of legitimacy and entices people to participate.

REGIME CHANGE
BEGINS AT HOME
DICK CHENEY
Vice-president of the United States
Former CEO Halliburton
"I have great affection and respect for Halliburton."
Cheney continues to receive over $1 million a year "deferred compensation" from the company.
TONY BLAIR
aka "Bomber Blair"
British prime minister
"It's worse than you think. I really do believe in it"
—on the neo-liberal project of New Labour.
HORST KÖHLER
Managing director of the IMF
DONALD RUMSFELD
US Secretary of Defense
"It has nothing to do with oil, literally nothing to do with oil."
But then again...
"In wartime, truth is so precious that she should always be attended by a bodyguard of lies"
(quoting Churchill).

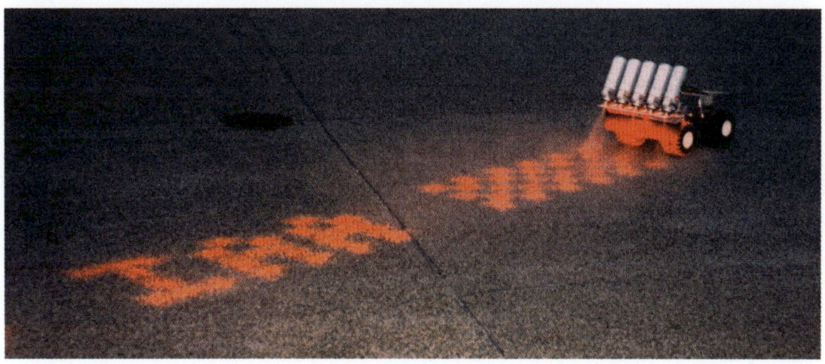

Top 'What's wrong with McDonald's? Everything they don't want you to know', pamphlet, London Greenpeace Group • Printed paper • London, 1986 • V&A: 38041006204655 • In 1986 a small group began handing out this leaflet outside McDonald's in London. McDonald's sued them for libel and they defended themselves in the longest civil trial in English history, massively amplifying the leaflet's impact. Around three million copies of a fresh version of the pamphlet were distributed in Britain and it was translated into 26 languages. The 'McLibel' leaflet marked an international watershed, after which corporations have been reluctant to use libel laws to silence dissent. In 2013, it was revealed that the pamphlet had been co-written by Bob Lambert, an undercover police officer who spent five years infiltrating the campaigners. The revelation is part of an ongoing scandal in Britain concerning the use of public resources for spying on peaceful protest groups and the conduct of undercover police.

Opposite Top Nike blanket, created by Cat Mazza and an international group of knit and crochet hobbyists • Knitted and crocheted wool, 2003–8 • Private collection • From 2003 to 2008, a diverse group of international craft hobbyists from over 30 countries participated in the Nike Blanket Petition project to create a 15-ft/5-m wide quilt of the Nike 'swoosh' logo. Each handmade square acts as a signature for fair labour practices and better conditions for Nike garment workers. The petitioners represent craft workers mobilizing against the exploitation of labour by global corporations.

Opposite Bottom The collective microRevolt created a series of 'logoknits' – knitted garments with corporate logos. These were intended to generate discussion on the relation between craft, labor, production and consumption, as well as digital copyright. Logoknits are made with patterns generated from microRevolt's free web application knitPro.

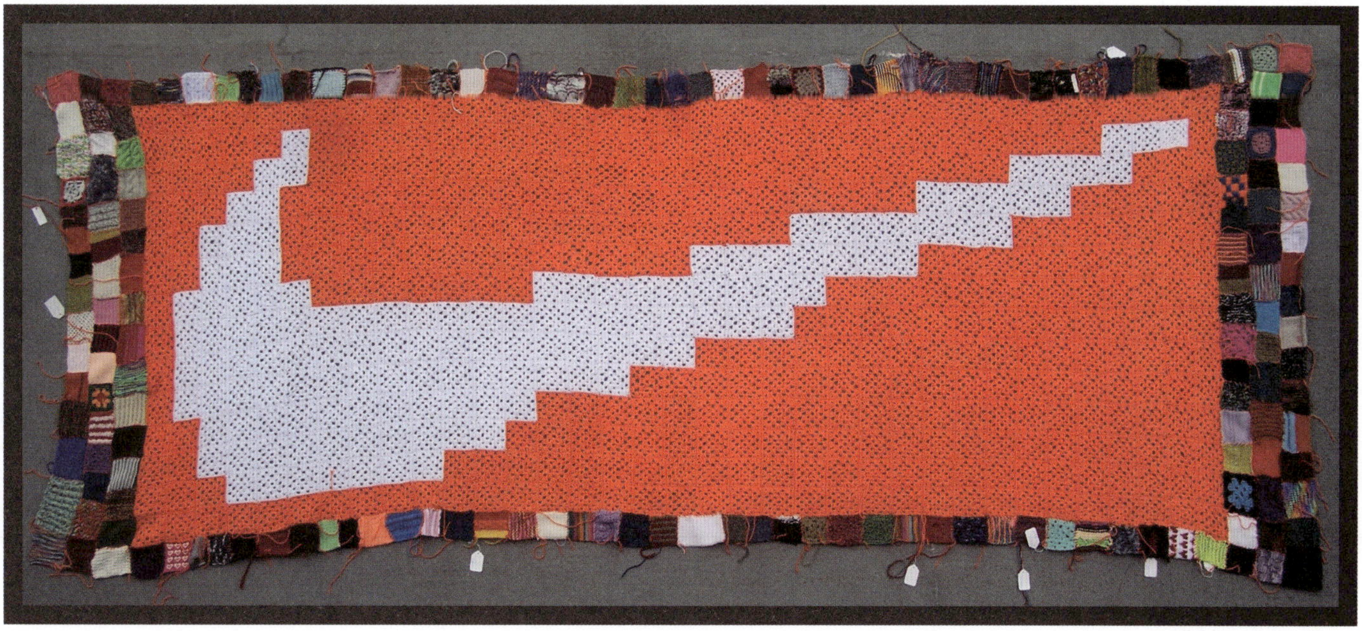

KNITTING PATTERN

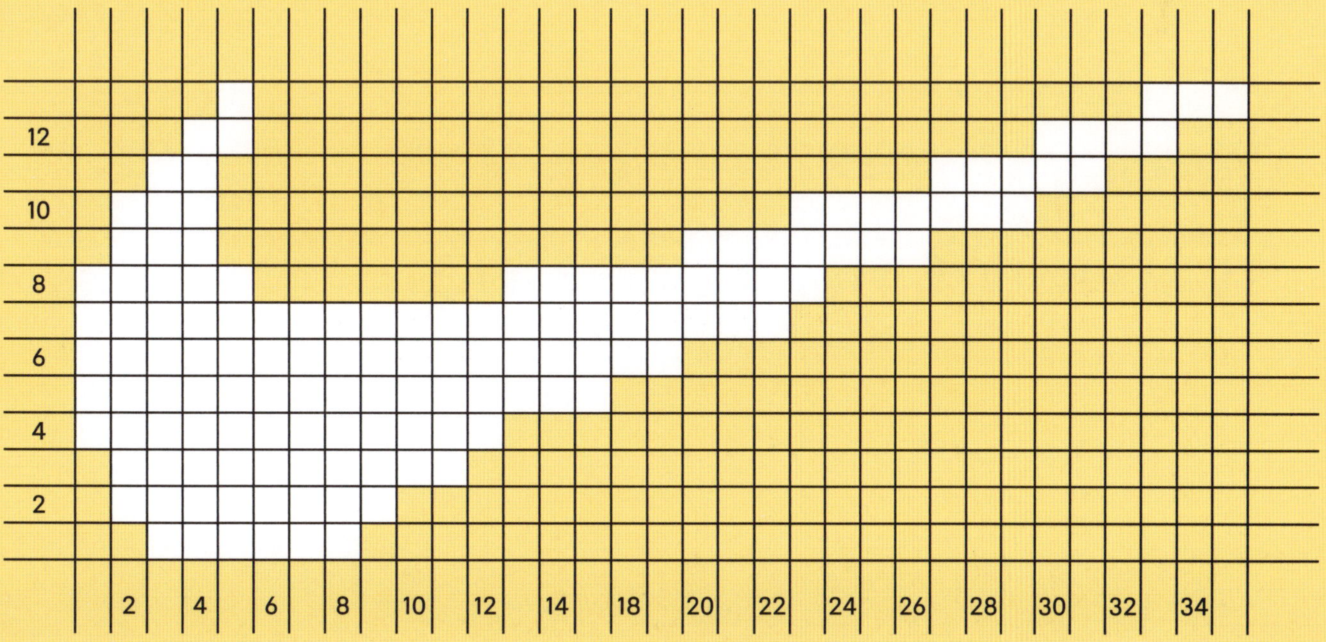

public billboards, paintings, walls, pavements, easels, trees, T-shirts, sweaters, banners, graffiti, videos, stencils, signs, surveys, maps, traffic signs, notices, school aprons, walls, monuments, trees, paper, silk-screens, stamps, murals, silhouettes, photocopied pictures, photographs, small placards, bodies, hands—

Disobedient Bodies: Art Activisms in Argentina

The linking of political action and art practices has a long history in Argentina, going back at least as far as the late nineteenth century. Far from unproblematic and peaceful, this combination has been marked by tensions, discrepancies, utopian propositions and secret alliances. Some of these episodes have been repeatedly revisited in the last few years, particularly the confluence of avant-garde and political radicalization that generated the frenzied *1968 Itinerary*,[1] and culminated in the collective work *Tucumán Arde.*

This overview focuses on those practices of 'artistic activism' that have developed in Argentina since the 1980s, concentrating on recent history and the current situation of some of the most active groups. The term 'artistic activism' is here employed to encompass mostly collective productions and actions that tap artistic resources with the goal of adopting a political stance in the public sphere.

Three phases, roughly corresponding to the last three decades, are critical to the emergence and proliferation of activist art practices linked to new social movements in Argentina.

THE EIGHTIES

On 24 March 1976, a new military coup, headed by General Jorge Rafael Videla, disrupted the fragile institutional life in Argentina, thus initiating the darkest period in the country's history. The military government planned and carried out a gruesome terrorist campaign to annihilate any type of opposition among the population, not only that coming from guerrilla groups, but also from dissident activists among trade unionists, students and the intelligentsia. Around 500 clandestine detention camps were created throughout the country, where thousands of illegally arrested people were brutally tortured and, in most cases, killed. The perverse figure of the 'disappeared', whose whereabouts were unknown from the moment they were kidnapped, began to paralyse the entire society as a result of that which Pilar Calveiro calls the spreading of 'concentrationary terror'.[2] In 1983, the dictatorship ended in a retreat action precipitated by the defeat in the Malvinas/Falklands War, leaving in its wake the grim balance sheet of approximately 30,000 disappeared, half a million exiled, a dismantled national industry, an exponentially increased foreign debt and other obvious signs of the implementation of a neo-liberal agenda in the region.

From the very beginning, the *Madres de Plaza de Mayo* adopted symbolic strategies that identified them, enhanced their cohesion as a group and made them and their cause visible – to the relatives of other disappeared people, to the rest of Argentine society, to the foreign press and to the international community. Among the different creative strategies used by the *Madres* since the beginning of the dictatorship to give visibility to the genocide, three different types of objects can be distinguished: photographs, silhouettes and graphic actions in the streets. Early on in their struggle, the *Madres* improvised small placards with the photos of their disappeared children and hung them from their bodies or held them up in their hands in their walks around the Plaza de Mayo – the symbolic centre of the capital – as well as in their visits to civil servants' offices. In this way, the *Madres* inaugurated a prolific

Right Placards with photographs of disappeared people, Buenos Aires, 1983.

genealogy: the photographs of the disappeared have become one of the most recurrent and powerful ways of remembering them. Those images reaffirm the existence of the disappeared as subjects with a biography that predated their kidnapping, an existence that was categorically

negated by the genocidal regime. Those who disappeared had a name, a face, an identity and a family who was looking for them and demanding their reappearance.

The widespread use of life-size silhouettes can be traced to 21 September 1983, when, still under the dictatorship, what has come to be known as *El Siluetazo* took place: the massive production of silhouettes meant to give physical presence to the 30,000 disappeared.[3] In the middle of a hostile and repressive atmosphere, a (temporary) space for collective creation was liberated – something that can be thought of as a redefinition of artistic and political practices. *El Siluetazo* was an event in the fullest sense of the word: one of those exceptional moments in history in which an artistic initiative spontaneously meets a demand coming from social movements and gains momentum, thanks to the irrepressible force of mass mobilization. The event involved a vast improvised outdoor workshop in which hundreds of demonstrators painted silhouettes of their bodies that were later pasted on walls, monuments and trees, despite the intimidating police presence that threateningly surrounded them. The silhouettes were made using the simple process employed to teach children to draw: tracing the outline of the human body on a piece of paper. The outlines thus obtained were meant to represent 'the presence of an absence' – the absence of thousands of disappeared. The key to this symbolic gesture was quantification: spatially signalling the void left among the population by thousands of disappeared bodies.

As to the graphic initiatives, the group GAS-TAR (later renamed CAPATACO)[4] took an active role in the artistic activism of this period, through actions developed in highly conflictual contexts in public spaces (squares, the streets, a factory on strike). From 1980, the group carried out a series of collective actions in Buenos Aires that included, among others, street silk-screen workshops for the production of graphic posters, the use of pavements as support for silk-screen stamps and murals made with photocopied pictures of the disappeared. The massive production of 'participatory posters' also invited people (demonstrators and passers-by alike) to complete the blank spaces in the image, e.g. the repressor's face or name.

In the group's graphic technique, the deviation from orthodoxy as a way to create a mixed and 'dirty' graphic tool available to everybody was evident. GAS-TAR daringly revisited pre-existing images

Top Young people painting silhouettes at the Obelisk of Buenos Aires during the second *Siluetazo* on the last day of the dictatorship, 8 December 1983.

Opposite 'FIM Porongovanguardia' poster, CAPATACO, designed by Emei, Fernando 'Coco' Bedoya and Joan Prim • Buenos Aires, 1984.

(from political propaganda or advertising) in order to subvert the symbolic charge of consensual discourse during the democratic transition. As an example of these interventions, we can see how in the poster 'FIM Porongovanguardia', 1984, from under the face of the recently elected president Raúl Alfonsín, as portrayed in his official image, a skull, as a reminder of death, threateningly emerges, while his hands, clasped together in his emblematic hailing gesture to the masses, wield a double-headed axe with the word FIM written on it, in a clear pun on the Spanish word 'fin' (end) and the acronym for the Fondo Monetario Internacional (International Monetary Fund).

THE NINETIES

The second moment takes place in the mid-1990s, a decade marked by the dismantling of the state and the vacuous ostentation of neo-liberal policy. In this context, there emerged a few isolated art groups that performed interventions in the streets as well as in art spaces. Two of these groups were the *Grupo de Arte Callejero (GAC*, Street Art Group)[5] and *Etcétera* (re-named the Errorist International in 2005), both closely linked to the creation of *HIJOS*[6] – the human rights organization established in 1999 to group the sons and daughters of the disappeared, many of whom had become the exiles and militants of the 1960s and 1970s. *HIJOS* in turn learned much from the *Madres de Plaza de Mayo*. But unlike the marches of the *Madres* that had taken place every Thursday since 1977 – which consisted of a walk around the Plaza de Mayo – the *escraches* (exposure protests),[7] in which both the *GAC* and *Etcétera* took an active role, functioned as a 'de-localized' and dispersed practice. They could take place at any time and wherever an unpunished criminal happened to live: '*Adonde vayan los iremos a encontrar*' ('Wherever they are, we'll find them') was a common chant in human rights demonstrations. In addition, while the *Madres* employed a symbolic strategy designed to give visibility to the disappeared, whose existence the dictatorship perversely refused to acknowledge, *HIJOS* concentrated its efforts instead on exposing the perpetrators of the genocide and seeking their popular indictment.

The *GAC* was founded by students and graduates of the Escuela Nacional de Bellas Artes Prilidiano Pueyrredón, who in early 1997 made unsigned murals by pasting traditional white school aprons on walls in support of a prolonged teachers' strike. The members of *Etcétera* defined themselves as self-taught Surrealists, linked to the underground theatre scene. Despite their different origins, the *GAC* and *Etcétera* worked side by side in many actions, especially as a result of their shared collaboration with *HIJOS* in the staging of *escraches*.

A typical scene: on a day like any other in the late 1990s, somewhere in Buenos Aires, a group of young people are plastering the streets with posters, while talking to the neighbours. The posters show the face of a man who lives in the neighbourhood, giving his exact address, phone number and record: he is an army officer involved in numerous cases of illegal detention, torture and disappearance. He was the head of one of the numerous clandestine

detention and extermination centres that existed in Argentina during the last dictatorship. He is free, thanks to the so-called Pardon Laws and amnesties granted by successive democratic administrations. The youngsters invite this man's neighbours to join in the exposure campaign to be carried out in front of the criminal's house some days later. '*Si no hay justicia, hay escrache*' ('If there is no justice, there is *escrache*') was the *HIJOS* battle cry.

In 1997 the *GAC* started producing the graphics for the *escraches*. Their typical notices subvert the traffic code by pretending to reproduce an ordinary traffic sign – in fact, such a sign might well remain unnoticed by many an unaware passer-by. But the signs are in fact indicating, for example, the proximity of what used to be a clandestine detention centre ('ESMA – 500 m. away'); or the names of the airfields from where the 'death flights' took off, carrying detainees to be dumped, while still alive, into the sea; or a demand for the prosecution and punishment of a 'repressor'. The collective's signs laid bare the enduring relevance of their cause: a quarter of a century later, those who had ordered and executed the genocide were still living among us, their identities and records all but forgotten. Copies of their anonymous cartographic work '*Aquí viven genocidas*' ('Criminals Live Here'), a map of Buenos Aires indicating the exact location of the homes of more than 100 individuals actively involved in the crimes against humanity committed during the so-called Dirty War, were plastered all around the city for the 25th anniversary of the 1976 military coup.

Top Street signs designed by *Grupo de Arte Callajero*, Buenos Aires, 2006. These road signs were first used on 19 March 1998, in conjunction with *HIJOS*. They indicate the homes of generals involved in the genocide and the sites of former genocides and clandestine detention centres.

Right '100 m to Cardinal Juan Carlos Aramburu, Complicit with the Military Dictatorship' street sign, designed by *Grupo de Arte Callajero* in conjunction with *HIJOS*, Buenos Aires, 2002. This sign was put up as part of the *escrache* protests during Holy Week in April 2002.

Opposite 'Genocides live here', *Grupo de Arte Callajero*, Buenos Aires, 2002. This map was distributed during the 24 March *escrache* protests. It shows the then-current addresses of generals complicit in the genocide, and a pamphlet distributed with it listed their phone numbers. It was reprinted with updates for protests in 2003, 2004 and 2006.

The *GAC* defines its production as a specific kind of militancy, 'a political militancy through art. We don't think politics necessarily needs to be

practiced through conventional means.'[8] Rather than taking politics as the subject, content or as an external reference, as in so-called 'political art', or serving as an agent for the aestheticization of politics, their work seeks to create 'a space where art and politics can be part of a single mechanism of production'. They produce ephemeral, multiple (usually anonymous) graphic resources and experiences (stencils, signs, surveys) that can be appropriated and circulated by others. The *GAC* are very clear and explicit about this: 'We rarely include signatures in our production. Most of our output is anonymous, which stresses the ambiguity of its origin. We encourage the reappropriation of our works and methods by other groups and individuals who share our general goals.'[9]

At the beginning, the *GAC*'s street signs and *Etcétera*'s performances that accompanied the *escraches* were invisible as 'art actions' within the art world, while at the same time providing the *escraches* with an identity and social visibility that helped them become a new and powerful form of struggle against impunity and injustice. The *escraches* contributed to the development of a new festive form of collective participation that made a subversive means of political intervention possible.

2001 AND AFTER

The third moment revolves around the unprecedented crisis and popular uprising of December 2001 in Argentina, with its aftermath of institutional instability and continuous unrest, in which new social movements played a leading role. Many collectives became involved in the broad call for substantial change in the political system – summarized in the radical slogan '*que se vayan todos*' ('out with them all'), addressed to the entire political leadership of the country. In the heat of the revolt, a significant number of visual artists, film-makers, poets, alternative journalists, thinkers and social activists invented new forms of intervention linked to social events and movements in the expectation that they would change life in Argentina: popular assemblies, pickets, 'recuperated' factories, movements of the unemployed, bartering clubs and so on. The subversive use of mass media and the development of alternative means of communication were tools common to the new forms of protest. Art groups quickly responded to the call of new collective organizations demanding a radical change in the political system, and were involved in the emergence of renewed forms of activism. In those days of intense social turmoil, the art groups were showered with requests from assemblies and pickets, and inundated with continuous calls to demonstrations. They went as far as to produce weekly actions. Several art collectives would collaborate in one single demonstration and artists who worked in more than one group had to run from one action to the other. Some of these collectives had an ephemeral existence, linked to a particular circumstance; others, such as the *Taller Popular de Serigrafía* (*TPS*, Silk-screen Printing Popular Workshop) and *Argentina Arde* (Argentina Burns), survived until not long ago.

The *TPS* originated from a specific request made by a popular assembly in February 2002. Soon they were producing posters to promote demonstrations and then, in a spontaneous and fortuitous way, they found themselves printing their serigraphic images on articles of clothing – T-shirts, sweaters, banners, 'whatever people wear, take off and offer in amorous demand'[10] – at demonstrations and political assemblies, particularly at those organized by the movement of the unemployed. Working with a repertoire that, in equal measure, plundered and paid homage to Russian Constructivism, Argentine conceptual artist Víctor Grippo and other distant references, the *TPS* tried to 'provide the struggle with an image that might serve to identify the time and place of the protest'.[11] They did so, on the basis of a one-on-one exchange in a protest context, from the hand that printed to the hand that offered a personal garment in a potlatch act. The effect was the opposite of 'uniforming' people; this action was that of a voluntary mark on the clothes and on the body. For each occasion, they generated a wide range of direct, even obvious images and phrases that reflected the cause and emotional tenor of each specific action.

> 'The struggle to end sexist oppression that focuses on destroying the cultural basis for such domination strengthens other liberation struggles. Individuals who fight for the eradication of sexism without struggles to end racism or classism undermine their own efforts. Individuals who fight for the eradication of racism or classism while supporting sexist oppression are helping to maintain the cultural basis of all forms of group oppression.' Bell Hooks, *Feminist Theory from Margin to Centre*, 1984

Argentina Arde, like the historical action of 1968 its name pays homage to,[12] was a counter-information project that was born in the heat of the 2001 crisis. It came about as a result of a general call, launched by Indymedia – Independent Media Center, a non-official information initiative created in 1999 during the Seattle anti-WTO protests – to everyone who was documenting the events to help counter the disinformation coming from the 'controlled media'. '*Nos mean y Clarín dice que llueve*' ('They piss on us and *Clarín* says it's raining') was a recurrent graffiti from those days referencing the news reported in Argentina's largest newspaper. *Argentina Arde* operated like a permanent assembly – one of the numerous neighbourhood assemblies that flourished during the crisis – mirroring their horizontal structure, lack of hierarchies and decision-making processes. The collective had hundreds of members, with different committees dedicated to video, photography, press and cultural activism.

COMMON FEATURES

The artistic activism that developed during the 2001 crisis included a wide variety of practices and devices, from conventional media exhibited in atypical spaces (easel paintings hanging from trees at a public plaza in support of female workers who had 'recuperated' a factory in 2003) to experimental performative actions and urban graphic interventions on different surfaces (pavements, walls, posters, clothing, banners), from wall paintings in the style of the traditional Mexican and Latin American murals to massive exhibitions at institutional venues. The repertoire of expressive means employed by these practices was also very comprehensive, drawing on different traditions of so-called 'political art', from references to social realism to the reactivation of the historical avant-garde, including a revisitation of the critical experiments of the 1960s. Most of these artists – usually grouped in collectives – brought their interventions to the public space, operating during demonstrations and using city walls, public billboards and the streets. They addressed casual audiences, primarily consisting of people who were unaware of the artistic nature of their actions, often eliciting their interest, humour or puzzlement. The subversive use of mass communication means – urban graphics, billboards, posters – along with the newer generation of alternative communication became common practice in these new modalities of protest. Also common were the attempts to radically reappropriate the public space through the socialization of art. The actions sought to engage the public, to invite participation in the work. In some cases, hundreds of people became collective art producers, taking an active role in the creation of transformed subjectivities by engaging their own bodies in the actions, as well as in the use and circulation of the produced images. Quite often the 'artistic' origin of the practice became blurred, or was obliterated, as the resources contributed by the collectives were appropriated and re-signified by the crowd. These initiatives developed through a process of critical rethinking and reformulation of the legitimacy of traditional forms of representation both in politics and art. The collectives also revised their own complex position within the institutional circuits of the art world, as well as within the social and political spheres.

DILEMMAS

Since Néstor Kirchner took office as President in 2003, political and economic stability and a hegemonic pact for governance have been re-established. In this new and complex scenario, social movements are disbanding, losing the impetus they once had and, in many cases, reverting to traditional political relations based on clientelism and parties. Among the new forms of activism, a deep rift has opened between those who support the government and those who oppose it. This divide alienates people who not long ago shared the same struggles in their fight for human rights policies – in particular the trial of the genocides – to be made a priority of the new government's official agenda.

Simultaneously, some art collectives also suffered a crisis due to the visibility and legitimacy that their practices gained on the international art circuit. The intensity and disrupting force of the Argentine uprising in 2001 caught the eye of intellectuals and activists, as well as artists and curators from all over the world, who saw in it a new and fertile socio-cultural laboratory. A number of activist art practices, which until then had remained stubbornly outside institutionalized art channels, now suddenly found themselves in international demand. A few groups were invited to participate in prestigious biennales and shows in Europe, America, Asia and even Oceania.

The over-exposure to which some Argentine activist art collectives have been subjected, both at a local and international level, has undoubtedly affected them, impacting on their practices, their ideas, the networks of relationships and affinities they had built – in short, the whole framework underlying their collective and individual subjectivities. Today, another global crisis having turned away from Argentina the international attention temporarily focused on it, the time is ripe for a revision of the consequences of that unexpected international over-exposure and its effect on these activist art practices.

The first group to undergo this experience was the *GAC*. Its most critical moment occurred in 2003, following an invitation to participate in the 50th Venice Biennale. The swift arc that projected its members from street activism to their inclusion in such prominent international art spaces generated undeniable tensions within the collective. These were finally resolved when they decided – after several more experiences and much discussion – not to show their productions in conventional exhibition spaces.

For them, the Venice Biennale represented €2,400 with which to fund their street activities. Other activists, in turn, knew quite well that inclusion in mainstream art circles did not come for free: there's always a price to pay. Brian Holmes summarizes this dilemma in the following terms:

> In the age of corporate patronage and the neoliberal state, art is becoming a field of extreme hypocrisy ... The temptation is then

> to cease playing the game (the anarchist solution), or to simply exploit the museum's resources for other ends ('radical media pragmatism'). Both positions are justified, from the activist point of view ... The most interesting question within the artistic field then becomes: How to play the exhibition game in such a way that something real can actually be won?[13]

In the case of the *TPS*, the consequences of this over-exposure were dramatic: in a few months between 2006 and 2007, they were invited to four biennales as well as major local and international exhibitions. This huge demand forced the group to concentrate solely on these events, disrupting their active links with social movements and finally precipitating their dissolution.

In recent years, as the need or demand for street demonstrations has decreased, some groups disbanded while others opted for introspection and withdrawal. Meanwhile, new forms of artistic activism have appeared: *Mujeres Públicas* (Public Women), a collective that took the cause of legal abortion and gender questions to the streets and altered common perceptions of the Feminist movement in Argentina; *Iconoclasistas*, whose complex cartographic studies circulate around the world;[14] *La Movida del Diablo* (The Devil Party), which was formed the day the universal marriage law was passed, making Argentina the first country in Latin America to legalize same-sex marriage; *Luli*, a group that carried out several graphic and urban interventions in the city of La Plata to denounce police repression; *Serigrafistas Queer* (Queer Serigraphers), a circumstantial collective gathered to print funny and perplexing slogans on T-shirts on the occasion of the LGBT pride demonstrations, and many more.

'The defiance of established authority, religious and secular, social and political, as a world-wide phenomenon may well one day be accounted the outstanding event of the last decade.' Hannah Arendt, *Crises of the Republic*, 1969

The history and influence of these art activist collectives is also manifest in the more recent practices of political groups and in popular culture at large. It is noteworthy that the 'creative dimension' has been incorporated into a wide range of social protests, and can easily be seen in the upsurge of spontaneous, anonymous graphic practices (stencils, signs, interventions on conventional advertisements, etc.) found everywhere.

The disobedient objects that have been produced by these activist practices ever since the dictatorship are ephemeral and circumstantial devices of a precarious materiality meant to generate actions in public space, and new ways of opposing impunity and injustice. These are not sophisticated elaborations based on a hermetic rhetoric, but rather easily appropriated resources, reproducible actions, common types of knowledge, and re-used ideas, sometimes trite and predictable. Their 'artistic' qualities in terms of originality, authorship and contemporary critical debate are of little or no relevance. Instead, their value lies in the new ways of life and of relating to others that they created, turning shortage, grief and wrath into something else – a colourful call to others in times of frantic social ebullience. They are a far cry from what an art piece should be according to traditional views, since they distance themselves from individual authorship and their works aren't signed, can be made by many people and are just there for whoever wants to appropriate and modify them. In most cases, these groups make use of the available resources as if they were no more than a toolbox ready to be reactivated by the particular demands imposed by each new context. The objects that result from these practices are signs that sneak into the street, in the middle of daily life, disturbing or subverting civil norms or calling people's attention to what has been silenced or negated. They were, and still are, effective triggers when it comes to giving visibility to an experience of resistance and protest. Therefore, the disobedient condition of these objects, and of the collective uses and practices they activate, impacts not only on the art world but also on politics and law.

CLANDESTINE SYRIAN STENCIL

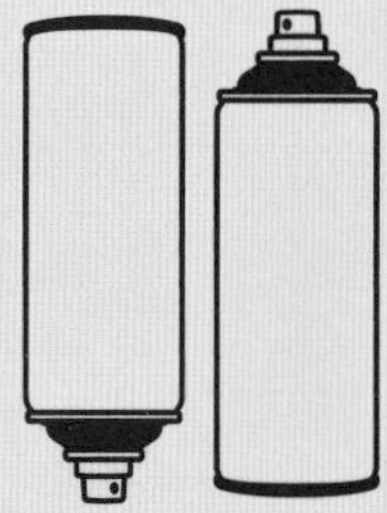

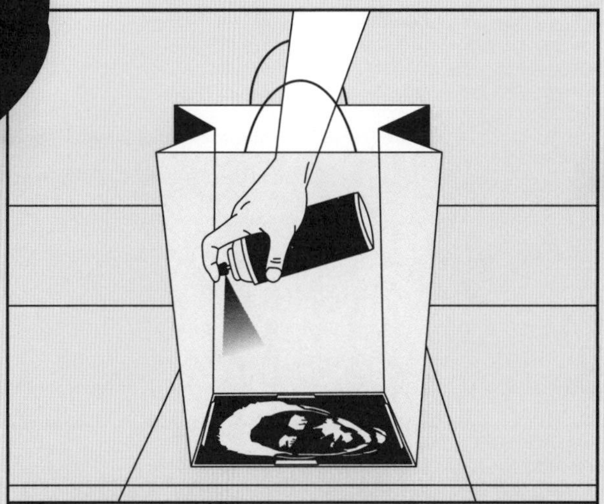

Paper Bag method: Cut out the bottom of a paper bag, leaving enough of the bottom intact to act as a frame. Tape the stencil to the bottom of the bag with clear packing tape to prevent paint bleeding through.

GHAITH MATTAR

Ghaith Mattar was a Syrian activist from Daraya much admired for his inspired ideas about non-violent protest. He was arrested in Damascus in September 2011 and died under torture. This stencil designed by Hasan Khzam is one of hundreds commemorating Syrian martyrs that are shared online and sprayed on walls in Syria and beyond.

SILENCE=DEATH
SILENCE=DEATH

Soli-dari-ty

‘If you come only to help me, you can go back home. But if you consider my struggle as part of your struggle for survival, then maybe we can work together.’ Aboriginal Activists Group, Queensland, attributed, 1970s

Previous ACT UP demonstration at Federal Plaza, New York City, 30 June 1987. Reproduced across placards, T-shirts and badges, the Silence = Death logo became synonymous with the ACT UP campaign to confront the AIDS crisis in the late 1980s. A pink triangle (pointing down) was the symbol used in Nazi concentration camps to mark out gay men. Appropriated (and inverted) as an organizing tool by the gay community, it made a powerful point about the need to challenge silence over the annihilation of gay people.

Opposite From 1994, indigenous women in Chiapas began representing leaders of the EZLN (Zapatista Army of National Liberation) in masked adaptations of Chamulita dolls, whose fabric and thread use pre-Columbian techniques of preparing wool from sheep, sacred in Tzotzil culture. The Zapatistas emerged on New Years' Eve 1993, seizing control of the colonial city of San Cristóbal de las Casa, but grew through non-violent political means and international support. Sold to visitors, the dolls are a means of symbolic and economic solidarity. These two examples, which represent Zapatista leader Trini, were collected in Chiapas during the 1996 Intercontinental Encuentros for Humanity and Against Neoliberalism. More than 3,000 grassroots activists from over 40 countries attended, agreeing to create an intercontinental network of resistance and communication, People's Global Action.

Top, Right During the 2012 Quebec student protests, which halted proposed rises in tuition fees, red felt squares – the symbol of student protest in 2005 – became popular. The squares were simple crafted items with a strong iconographic element reproducible in other media. In 2013 a red feather was adopted to signify solidarity with the indigenous Idle No More movement.

Top Left 'Fuck the Law' pendant, commissioned by Herman Wallace • Chrome-plated steel • Louisiana State Penitentiary, United States, 2008 • Private collection • Herman Wallace was one of the 'Angola 3', three inmates of Louisiana State Penitentiary (Angola Prison) who in 1971 established a chapter of the Black Panther Party in the prison organizing prisoners against the brutal conditions there. They have served extraordinary periods of solitary confinement – over 100 years between them. Wallace died in 2013 after 41 years in solitary confinement, a few days after his conviction had been overturned and the judge ordered his immediate release. Robert Hillary King was released in 2001 and currently Albert Woodfox remains in prison. The men believe that it is because of their political convictions that they have been treated so harshly. They have been the subject of an international campaign protesting both the validity of their convictions and the inhumanity of their continuous solitary confinement. Herman Wallace got a fellow inmate to make this pendant for Poppy Newell-Richards, a British supporter who visited him in prison when she was eleven years old.

Top Right Zu-Tag pendant, design by Kenny Zulu Whitmore • Chrome-plated steel • Louisiana State Penitentiary, United States, 2013 • Private collection • Kenny Zulu Whitmore joined the Black Panther Party in prison where he met the Angola 3 and became politicized. He remains in Angola Prison, where he has been in solitary confinement for over 35 years. From his cell he encourages others in their political struggles, and through letter-writing he has gained friends all over the world. He designs pendants, which are handcrafted by other prisoners from basic materials and sold to generate funds for his legal campaign. People wear them to express solidarity with his cause.

Opposite Badges of the Struggle Against Apartheid • *c*.1980–94 • International Institute of Social History, private collections • These badges oppose apartheid, a system of racial segregation initiated by the all-white National Party in South Africa in 1946. These are just a fraction of the badges created and worn in the struggle. **Top two rows:** produced inside South Africa. Periods of emergency rule in the 1980s saw homes raided by police and activists arrested without charge. These badges mark how they organized. **Third row:** made by liberation groups forced underground and into exile by the National Party. **Rows four to nine:** produced and worn in solidarity, in countries from the UK to communist Bulgaria. Some organizations worked with exiles, but most who wore these badges never lived in South Africa. They supported the struggle by advocating for divestment, cultural and economic boycotts, sanctions and racial equality. **Last two badges** (*c*.1990s): 'Mandela Libre, La Lutte Continue' ('Mandela free, the struggle continues') and 'Mandela for president' mark a culmination of the movement through the transition to South Africa's first democratic elections.

NUSAS IS STUDENT POWER
HANDS OFF COSATU
LESBIANS AND GAYS AGAINST APARTHEID
UDF UNITES! APARTHEID DIVIDES!
resist forced removals
Viva Cosatu WORKERS UNITE !
Educate, Consolidate and Advance to Victory!
WOMEN MAKE HISTORY FIGHT FOR PEACE
WE SALUTE THE WOMEN
ARTISTS OF SOUTH AFRICA UNITE
AZASO EDUCATION TOWARDS DEMOCRACY
UNITED DEMOCRATIC FRONT MAY DAY
computers for people's power
S.A. CONGRESS OF TRADE UNIONS
African National Congress
AFRICAN NATIONAL CONGRESS WOMEN FIGHT FOR FREEDOM
ANC YOUTH SECTION
Free Nelson Mandela
SAVE THE A.N.C.6
S.A.C.P. SOUTH AFRICAN COMMUNIST PARTY
OMROEP VOOR RADIO FREEDOM
Children of AZANIA unite
СВОБОДА ЗА НЕЛСЪН МАНДЕЛА
FREE NELSON MANDELA!
END MILITARY COLLABORATION WITH APARTHEID
MANDELA LIBRE !
FREIHEIT für MANDELA · Südafrika ·
LIBERTE POUR JAMES MANGE
NO TO APARTHEID solidarity with black workers
SUPPORT S.A. WAR RESISTERS AGAINST APARTHEID
DEFEND SOCIALISM AT BIRTH
Apartheid Runs On Fluor Power
BEAT! APARTHEID
BIKO LIVES
City of London Anti-Apartheid Group March for Mayekiso 10 Oct '87
Give Botha a necklace!
REMEMBER JUNE 16, 1976 FREE SOUTH AFRICA
ARTISTS AGAINST APARTHEID
UNITED NATIONS WORLD AGAINST APARTHEID
Freiheit für Namibia und Südafrika !
ORGANIZATION
APARTHEID NEIN
FIGHT APARTHEID NO TOUR '85
BOYCOTT SYDAFRIKA
BLIJF VAN M'N MAKKERS!
END APARTHEID TERROR! SANCTIONS AGAINST SOUTH AFRICA!
ISOLERA APARTHEID
BOYCOT APARTHEID
DISINVEST BREAK THE SOUTH AFRICAN CONNECTION
STRIKE AGAINST APARTHEID for black workers power
APARTHEID NO SANCTIONS YES
Stoppt die atomare Zusammenarbeit Bundesrepublik – Südafrika
STOP OIL SUPPLIES TO SOUTH AFRICA
BOYCOTT PRODUCTS OF APARTHEID
STOP OLIE naar zuid-afrika
WOMEN AGAINST APARTHEID
LIBERTE POUR L'AFRIQUE AUSTRALE mrap
MANDELA LIBRE LA LUTTE CONTINUE RNCA
MANDELA FOR PRESIDENT THE PEOPLE'S CHOICE

Below 'Solidarność' badge, anonymous (logo design by Jerzy Janiszewski) • Lithograph and screen print on polystyrene; dye-stamped metal and polystyrene • Poland, 1980–8 • V&A: E.2255–1990 • Axial-lead resistor • plastic and nickel-chrome wire • 2014 • Private collection • 'Solidarność' (Solidarity), founded in September 1980, was the first free trade union in an Eastern Bloc country and became a leading political force in opposing communism. In December 1981 martial law was imposed in Poland in a crackdown on Solidarność, which was declared illegal. Supporters wore tiny badges with the Solidarność logo, which signalled their support for the movement in a way that could be easily concealed. A more oblique strategy was to attach a 'moc rezystor' (power resistor), taken from a domestic radio, to your lapel – a play on words which indicated resistance to the government and support for pirate Radio Solidarity.

Opposite This placard, shown in use in Tahrir Square, Cairo, was posted online in October 2011 by Egyptian activists as an expression of solidarity with America's Occupy Oakland protesters following a violent attack, days before, by the Oakland police department on Occupy protesters who refused to go home.

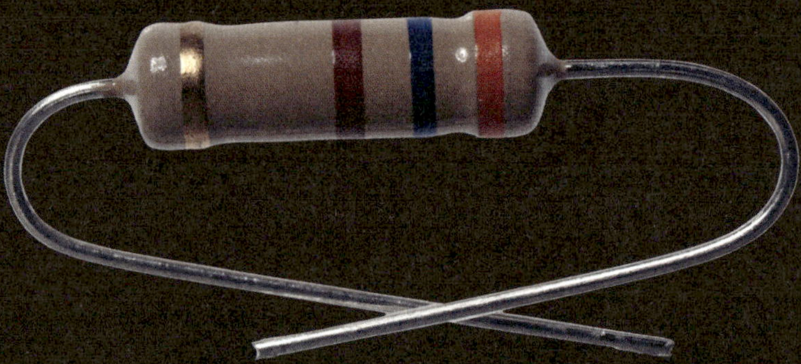

BUCKET PAMPHLET BOMB

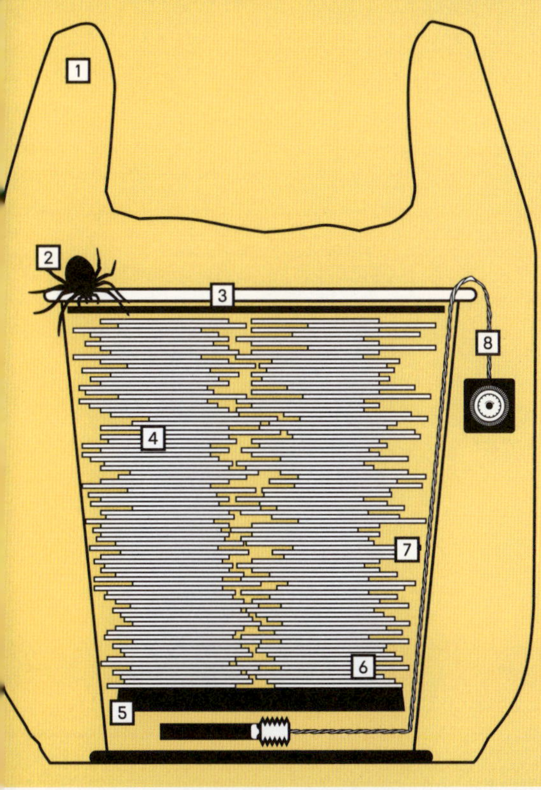

1 Carrier bag
2 Toy spider to deter inquisitive people
3 Warning notice
4 Pamphlets
5 Small explosive charge
6 Wooden platform
7 Pair of wires
8 Time switch

Port Elizabeth EVENING NEWS, Friday, August 14, 1970 HH

LEAFLET BOMBS WARN VORSTER: END IS NEAR

From DOUGLAS ALEXANDER

JOHANNESBURG.

TWO BOMBS exploded in Johannesburg today scattering hundreds of leaflets attacking South Africa's government.

One bomb exploded in Diagonal Street, a non-white area.

Pamphlet bomb used by the London Recruits, mostly young non-South Africans voluntarily working for the ANC and SACP. The pamphlet bombs harmed no one, but distributed hundreds of pamphlets high into the air, circumventing apartheid's censorship in cities across South Africa from 1969 onwards. The devices were invented and tested by ANC exiles in Britain. Several Latin American leftist guerilla groups also used them from 1980 onwards.

A Multitude of Struggles

'There is no such thing as a single-issue struggle because we do not live single-issue lives.' Audre Lorde, *Sister Outsider*, 1984

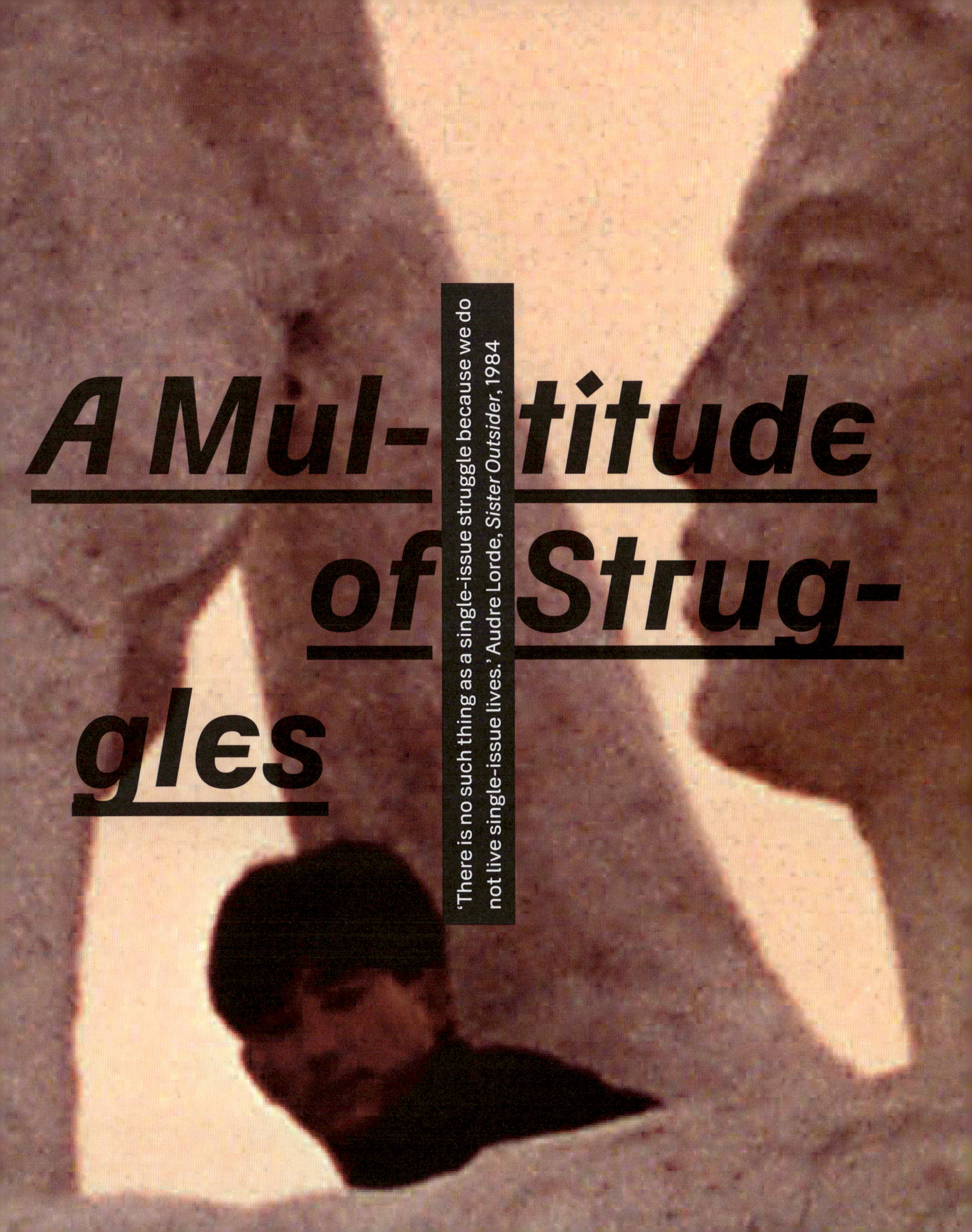

THE CLOCKTOWER

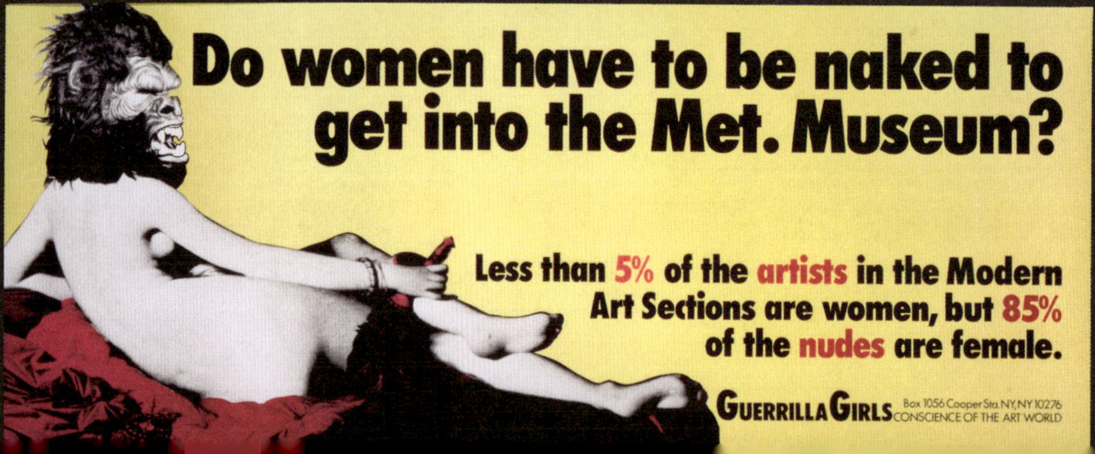
Do women have to be naked to get into the Met. Museum?
Less than 5% of the artists in the Modern Art Sections are women, but 85% of the nudes are female.
GUERRILLA GIRLS
Box 1056 Cooper Sta. NY, NY 10276
CONSCIENCE OF THE ART WORLD

Previous This 30-ft/10m statue of foam and papier-mâché on a steel frame was constructed by protesters in Tiananmen Square, Beijing, in May 1989, reinvigorating the protests there. It was based on Soviet sculptor Vera Mukhina's *Kolkhoz Woman* of 1937. Smuggled into the square, it dramatically faced Mao's portrait. It was destroyed five days later by a tank in the government assault on the protesters that ended the Pro-Democracy movement. Replicas around the world recall it, including the 'TSquare' augmented reality layer, which, on a mobile phone screen, projected the statue back into the square.

Opposite Guerrilla Girls photographed with fly-posting equipment in New York in 1991 • 'Do women have to be naked to get into the Met. Museum?' poster, Guerrilla Girls • Print on paper • United States, 1989 • V&A: E.622–997 • In 1985 a group of women artists founded the Guerrilla Girls to expose sexism, racism and corruption in the art world. They achieve this with graphics and public actions that deliver facts with humour. The members of the Guerrilla Girls maintain anonymity by wearing gorilla masks (describing themselves as 'feminist masked avengers in the tradition of anonymous do-gooders like Robin Hood, Wonder Woman and Batman'), a tactic intended to keep the focus on the issues rather than their personalities.

Top Puppets and puppeteers from *Top Goon: Diaries of a Little Dictator*, performed by the Syrian artist group Masasit Mati, 2011–12. *Top Goon* is a web-based series that lampoons Syrian President Bashar al-Assad and his regime's response to the popular uprising. By using finger puppets Masasit Mati are able to shield their identities while reducing the president to a figure of fun.

Right This protest palanquin reworks a Japanese Mikoshi, a traditional religious transport for deities used in festivals, adding subcultural motorcycle decorations. Its maker, Muneteru Ujino, found Mikoshis received less attention from the police than wheeled vehicles. This one was made for and used in a 2003 demonstration in Tokyo against the invasion of Iraq. Here, artist-activist Masanori Oda rides on it with one of his many DIY noise instruments made especially for protests.

***Arpilleras* (pronounced ar-pee-air-ahs) are three-dimensional appliquéd textiles, originally made by women in Chile to document the violence and repression suffered under the Pinochet dictatorship. Through *arpilleras* women found strength and solace by gathering together in workshops. They challenged their poverty by producing a grassroots export that could be sold abroad, and they were able to express the realities of their lives through stitches. *Arpilleras* record atrocities inflicted by the military and the everyday economic struggle to survive. They also proudly tell the story of women's acts of protest and resistance.**

***Arpilleras* are made from basic resources. Empty flour sacks provide the hessian backing on to which are stitched scraps of material, sometimes from the clothes of loved ones taken away by the regime. Dismissed as folk art, *arpilleras* existed for a time below the radar of political censorship, raising awareness and strengthening solidarity outside Chile. Now they are living memories of situations that it was not permitted to speak of.**

Bottom '*¿Dónde están nuestros hijos?*' ('Where are our children?') *arpillera* (front and back), anonymous • Appliquéd fabric scraps, Santiago, Chile, 1979 • Roberta Bacic Collection • A subject of many *arpilleras* is that of the disappeared, people detained by the military junta in Chile who vanished without trace. Many families still do not know the fate of their loved ones, or where they are buried. This *arpillera* shows a weeping mother with symbols of chained fists and doves falling from the sky. As can be seen on the reverse of this piece, women sometimes stitched small pockets on to the back of *arpilleras*, which held tiny handwritten notes addressed to whoever bought the textile.

Opposite '*Desplazamiento*' ('Displacement') *arpillera*, *Mujeres tejiendo sueños y sabores de paz* (Women knitting dreams and the taste of peace) • Mampuján, Colombia, 2010 • Roberta Bacic Collection • The process of women depicting stories of adversity in the form of *arpilleras* has spread beyond Chile to other Latin American countries, Africa and Europe. This *arpillera* was created by a group of 15 women who survived a massacre in Mampuján, a small town in north-west Colombia, on 11 March 2000. More than 1,400 people from the area were displaced at this time. The scene depicts the military attacking the inhabitants, a home set on fire and people abandoning the town carrying their children in their arms. An old woman, too frail to walk, is borne in a hammock.

DESPLAZAMIENTO
EN COLOMBIA

TO KNOW
BUT NOT TO DO, IS NOT
TO KNOW AT ALL

Opposite The Tiki Love Truck keeps alive the memory of John Joe 'Ash' Amador, who was executed by the State of Texas on 29 August 2007, and incorporates his death mask. British artist Carrie Reichardt was commissioned by Walk the Plank to make the truck for the first Art Car Parade in Britain, which was held in Manchester, 2007. While she was working on it she received news that her friend Ash, with whom she had corresponded for a number of years, was going to be executed. She travelled to Texas to be a witness. Straight afterwards she and Ash's wife brought his body to a log cabin in the woods where Carrie's artistic collaborator, Nick Reynolds, cast the resin death mask. The making of the mask allowed Ash's friends and family to spend time with his body and provided a positive focus through the ordeal of the execution. Ten days later, back in Britain, the Tiki Love Truck, with Ash's mask in pride of place, made its way through the streets of Manchester with thousands watching. It has continued to tour festivals and parades as a spectacular statement against the death penalty. The Tiki Love Truck was designed by Carrie Reichardt and Thayen Rich. It was constructed by Reichardt and the Treatment Rooms Collective: Lori Bell, Linda Griffiths and Mark Wydler.

Below
Roy's Handkerchief, *Colectivo Bordamos por La Paz*, Mexico, 16 June 2013. *Bordamos por La Paz* groups embroider and display handkerchiefs weekly across Mexico to protest murders and forced disappearances. They honour the victims and advocate for justice and recognition. This handkerchief was embroidered by the mother of Roy Rivera, who was abducted at the age of 18. She paid a ransom, but he was never heard from again. Mexico's Interior Ministry reported that of over 26,000 disappearances from 2006 to 2012, only 5,319 appear in the government's official registry. Together these handkerchiefs question Mexican law enforcement and the government's promise to keep its citizens safe.

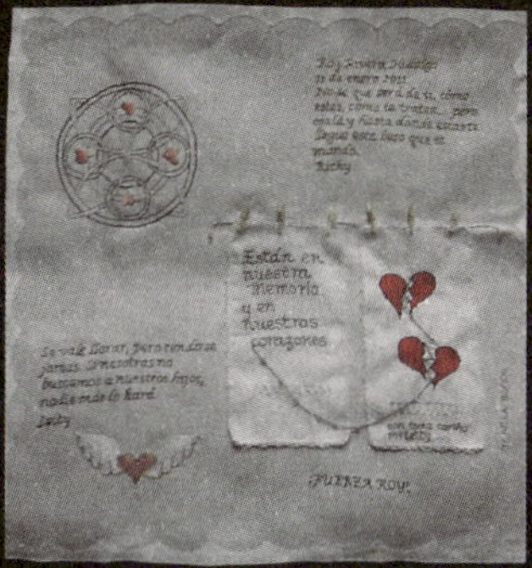

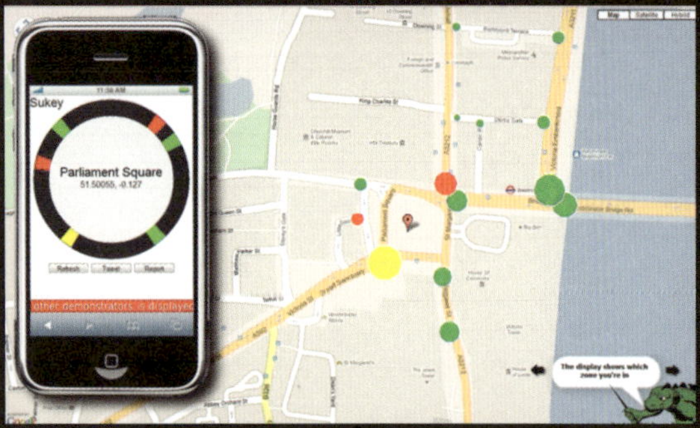

Top Left Sukey was a British organization, app and website developed in January 2011 to improve communications within student demonstrations. It was intended to provide realtime data on police and protester behaviour (sourced from SMS updates, TV, radio and social media) accessible in various formats to street protesters using mobile phones. Its name follows the English folk rhyme 'Polly put the kettle on, Sukey take it off again.' British Police have controversially used 'kettling' to trap and detain large peaceful crowds for as long as nine hours without food, water or toilets. Some critics allege kettling is used to provoke disorder and change the focus of public debate away from real issues.

Top Right Phone Story computer game, by Molleindustria, 2011. Phone Story is a game for smartphones that provokes awareness about troubling aspects of the manufacture and supply of the devices. Players have to force children to mine coltan in the Congo, prevent worker protest-suicides in a factory in China, manage rabid consumers in the West and dispose of electronic waste in Pakistan. Phone Story was banned from Apple's iTunes store four days after it was released in 2011.

Opposite Top The Tactical Ice Cream Unit, devised by the Center for Tactical Magic, was deployed at Occupy Wall Street, New York, in 2011, offering ice creams and pamphlets with a range of complementary flavours and issues, as well as gas masks and other protest supplies.

Opposite Bottom Bike Bloc, Laboratory of Insurrectionary Imagination/ Climate Camp, Hamburg, July 2010. The Bike Bloc was initiated by Climate Camp and the Laboratory of Insurrectionary Imagination in 2009 as part of the Reclaim Power protests during the 15th Climate Summit in Copenhagen. Discarded bikes were collected and re-welded with various modifications. One group, the Sound Swarm, comprised five bikes, each broadcasting part of a five-channel sound piece, improvisationally responding to different street situations. Others, Double-Double-Troubles, offered the facility to transport food, projectors or compost toilets to support a mass mobilization. On the day of action, over 200 bikes formed swarms, blockades and decoys to support thousands conducting civil disobedience on foot to breach the summit's security cordon and hold an alternative People's Assembly. NGO and indigenous peoples' representatives left the main summit to attend the assembly instead. The project was re-worked in Hamburg in 2010.

ICE CREAM

BIKE BLOC

1

Remove wheels, chain and brakes from the top bike. Cut off the seat stay and chain. Remove handlebars and seat post from bottom bike.

2

Sit one bike on top of the other and weld them together. Add extra material to strengthen, and join the forks together so the top bike steers the bottom. Add a new chain and brakes so the top bike drives the bottom one. Repeat with two more bikes. The pair of top bikes and the pair of bottom bikes should match as closely as possible.

3

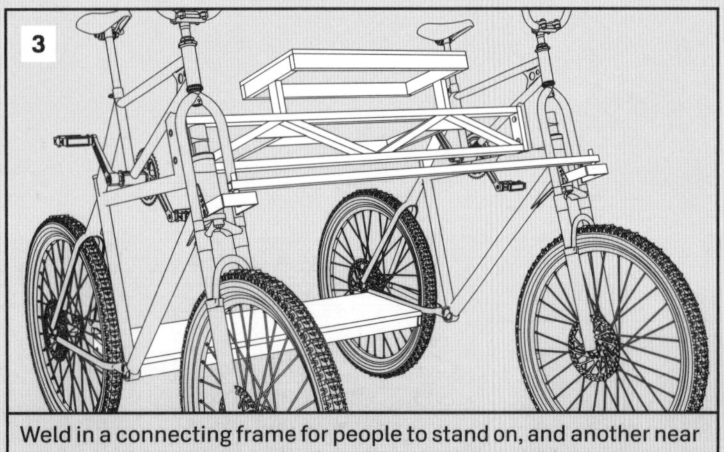

Weld in a connecting frame for people to stand on, and another near the front of the bike to make the entire unit into a rigid structure.

4

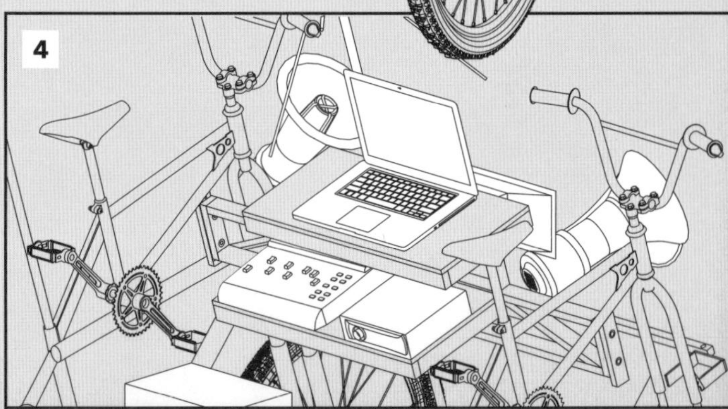

Customize! Speakers, laptops, batteries, mixers and FM transmitters make this bicycle machine into a performance space and pirate radio hub for taking to the streets.

This discussion was carried out over email in Autumn 2013, as the exhibition was being prepared.

John Holloway

These objects speak of pain and fury and dignity. They scream their 'No!' at us.

If our conversation is to have meaning, it must be to make that 'No!' louder, to magnify the pain and the fury and the dignity. There is always a danger that in placing an object in a museum, you silence it, you literally dumb it down. You, Gavin, have asked us to accompany the objects, with the intention, I hope, of getting us to turn up the volume, so that the visitor to this exhibition does not go out saying just 'how interesting, that was fun!' but leaves with rage in his or her heart. Rage because we too, all of us, are battered and beaten and trampled upon, and humiliated by the power of money. Rage because the obscenity of capitalism grows all the time: the grotesque inequalities; the lying, corrupt humbug of the politicians; the tearing up of the earth by the mining companies; the subjection of information, education and thought itself to the logic of a system based on profit. Rage because capitalism attacks us and attacks us and attacks us.

Look at that homemade gas mask over there, worn by the protesters in Gezi Park in Istanbul just a few months ago, and remember. Remember the simplicity of their demand. They wanted to stop a shopping mall being built on top of a park in the centre of the city, and for that they were beaten and gassed. And so they made masks to protect themselves and went on protesting. More than that, they turned the masks into a mockery of the system – '*chapuling*', they call it. There is a photo of the protesters dancing the tango in the park, with gas masks on their faces. Absolute mockery of the system, absolute proclamation that our struggle is not symmetrical to capital's struggle, that the world we want to build is quite different from the world of money. A beautiful echo, too, of the gas masks worn by the musicians in the centre of Sintagma Square in Athens just two years earlier, playing their bouzoukis as the gas grenades exploded around them.

Or look at that banner recalling the great miners' strike in Britain, and think of the women and men who fought and fought and fought not just for their jobs, not just for their traditions and their way of life, but against the surrender of the country to the great god of Money. Look at the banner, look around you at what the country has become and thank the miners for the anger and the dignity that they have handed on to us.

Or see the jewellery over there, made by the Black Panthers held in the prisons of the United States of Incarceration. See the pain turned into a thing of beauty, but a thing of beauty that does not and should not let us forget the pain that engendered it. Their pain, our pain.

Over there, that pot from the *cacerolazo* in Argentina tells us of the people who went down into the streets when their anger boiled over on 19 December 2001, who went out and chanted '*que se vayan todos*', 'out with the lot of them' – away with the whole miserable gang of politicians and exploiters. And they overthrew one president after another in the weeks that followed and showed the world how we can make it a different place, by taking over factories and hospitals, creating neighbourhood assemblies, setting up workshops and kitchens and schools. Not enough, but a blazing light of inspiration in the sky.

Look, then, at all these objects and feel the rage and dignity and hope they express, and know that that hope and dignity and rage are ours. And then, curator, commentators (us) and visitors, let us go out and shout it from the rooftops.

T.V. Reed

John has raised a number of key points. Most importantly, he has eloquently started to carry out the amplification these objects deserve, and has called upon us to remember the terror of the systems these decontextualized containers of rage resist. It is inevitable in placing these objects in a museum that their material and emotional context is obscured. Floating free from their home in political contestation, including life and death struggles, they are indeed in danger of becoming curios and fetishes. The surrounding texts and videos can certainly help reconstruct their original grounds, but a profound act of sympathetic imagination is also required. Cut out from the pain, suffering, but also the joy of struggle, they are like characters in a play that have wandered off stage. Only those who know the play can fully catch their meaning.

Of course, my choice of analogy is not random. Protest is always theatrical, and it is most effective when it is most imaginatively theatrical. This is not to trivialize, because all political life is theatrical. It is just that mostly all we see is reruns, both from the powers that be and all too often from resisters. My chant these days is 'A slogan/ exhausted/should never be repeated.' The current, rather jaded and bloated, mediascape calls for ever more artfully crafted acts of challenge to the systems that be. If there was ever a time when the arts of protest needed to be honed, it is now. In that sense, having these objects in a 'center for art detention' (Ishmael Reed) is not so inappropriate. If the lived context is dulled, the new context can serve as a reminder that creativity is absolutely key to meaningful transformation.

As I suggested, John is absolutely right to point toward the great pain and suffering that helped give birth to these objects. But these objects also contain great hope. Pain and suffering most often lead not to action but despair. The bravery of the producers of these objects is that they raged not inwardly but on the world stage, and those acts are acts of hope. That the symbolically real gas masks were featured in dancing reminds us that this is not only a mocking of the system, but also an embodiment of joy (Yes, Emma, fuck Lenin if he won't let you dance). One of the best-kept secrets about protest is that it is exhilarating. Throwing your whole being, body and soul, into a call for radical change is a profoundly positive, empowering act. Hannah Arendt wrote about 'public happiness', SNCC workers in the most deadly dangerous depths of the struggle against US racism and apartheid spoke of a 'freedom high'. Like all highs, freedom highs can be abused, can even become addictive, but these autonomous zones, however temporary, can have extraordinarily joyful power. They say 'No!' but they also say a great 'Yes!' to life lived intensely and meaningfully.

Julia Bryan-Wilson

I want to expand on the ideas of negation and exhaustion that were so eloquently articulated by both John and T.V., and connect them, however provisionally, to the ambivalence of repetition. I am struck by the fact that a hand-stitched banner, first created for a political occasion, Occupy LSX, is being remade specially for the exhibition at the V&A. One could say that this recreation, fabricated anew not for protest but for display, is a fraudulent duplicate, a bad copy that violates the object's intentions (even if it will be returned to its makers after the show). Why insist upon this *material* recreation? Why not show photographs of the banner in situ, with hands holding it aloft, as a reminder, precisely, of both its collaborative function and its loss? What sort of logic of substitution is being enacted when handcrafted objects such as banners are remade?

If a 'fine arts' piece such as a painting had been destroyed, would it seem normal for a museum to simply repaint it some years after it had been produced? Maybe, but maybe not: craft objects, especially textile-based works that are often collectively made (not least those created for protest contexts) lend themselves to replication because they are usually constructed out of easily located resources and were anonymously made in the first place. (Not to mention, of course, that in the West such textiles are associated with lesser-valued 'women's work'.) This repetition could be viewed as showing a dangerously blithe disregard for historical circumstance, replacing the textures of use and wear that such a banner accrues in its life on the street with a fresh version that might *look* the same, but holds none of that memory in its threads.

That's one side of the story: the suspicious side, the anti-institutional side, the side that believes that to 'museumify' is to deaden and de-fang, full stop. But repetition does not have to be understood only as an accomplice to regression: it can also be a potent reactivation. To return us to the images of gas masks, consider artist Allison Smith's recreations of handmade gas masks from the early part of the twentieth century in her 2009 project *Needle Work*.[1] These are based on her research at military history museums in the US and Europe, and in remaking them in all their strangeness, fragility and cloth flimsiness, she highlighted their apparent *insufficiency* as devices of protection. This is a queer repetition that forcibly drags the past into the present. Smith's masks disobey our tendency to think that current crises are somehow unique.

Smith also demonstrates that museums can function as usable archives, not just as repositories of prized things but also as holders of collective memory. Of course, we should be cautious about what happens when objects such as protest banners enter institutions divorced from their use, but that is true of all objects. And if not in museums, where would we rather have them? Cosseted in private collections? Mouldering in basements? At the end of the day, I applaud the fact that the 'Capitalism is Crisis' banner is being given a second life, along with a wider set of viewers. Let's acknowledge that an exhibition such as *Disobedient Objects* might be both a compromise and a revelation. Let's hope that the Museum itself becomes disobedient, fostering new kinds of conversations around these objects and performances and histories, and reaches out in radical ways to new publics. Alongside the calls for a more searing 'No!' and a more joyous 'Yes!', I say: Maybe. Sometimes. It depends.

Gavin Grindon

Julia opens up the question of the potential failure of the show, which is a real possibility, but she then leads us to ask on whose terms might it fail? Those of a newspaper critic? Those of the movement participant-researchers who helped establish its criteria and form? Another banner in the show comes from Russian protests against the government's election fraud and its incarceration of activists in the prison-industrial complex. Playing on the double-meaning of *predstavliat*, it reads both 'You don't even represent us' and 'You cannot even imagine us.' Resituated in this exhibition, it might be seen to also resonate against the exhibitionary complex (as Tony Bennet called it) and its limited ability to represent movement cultures. But should it be able to? In 2007 the Turbulence collective, in a paper distributed at the Heiligendamm protests, asked 'What Would it Mean to Win?'[2] Herman Wallace and Kenny Zulu Whitmore's jewellery, which John mentions, certainly resonates with these issues of ambiguous success and failure. We'd hoped to include a letter from Herman Wallace, addressed to the Museum visitor. But just last Tuesday, Herman had his conviction overturned and was released. Suffering from terminal cancer, he died three days later – a free man. Showing the jewellery he made now suddenly has very different resonances, but maybe that's no less true of any of these objects, whose meaning isn't resolved.

Jack Halberstam

Can there be a collectivity of objects? How do things live together in a public museum, in an exhibit, in a show? And how differently do they live together there than in the street, the house, the private gallery? Do the objects on display here only represent a disobedience that was performed elsewhere, or can there be a disobedience that emerges from their juxtaposition? When we call an object 'disobedient', do we mean that it captures and frames a disobedient gesture from another time and place or that it is disobedient to its status as an object and disobedient in its relation to the propulsive and willed function of the subject? Can the object refuse to be collected, fail to cohere, renege upon its signifying function?

These objects excite us in their multiplicity, their repetition and their implied use value. But the disobedience of objects might lie as much in their failure to capture or recapture the original context out of which they emerged (a protest, an occupation, a sewing circle, a riot, an intervention, an act of piracy) as it does in their ability to form a new vision of protest as they leave that context. What does the suffragette teapot say about gender, freedom and democracy when placed alongside a *cacerola*? How do the Syrian finger puppets of *Top Goon* signify when situated next to the Barbie Liberation Organisation's repurposed Barbie dolls? How does the Black Panther jewellery read alongside the anti-apartheid badges? The objects make visible lines of connection and solidarity between struggles as much as they offer images of the distinctness of each instance of protest.

When does a collection become a collective, unified less by theme and aesthetic value and more by intent and a shared sense of purpose and will? Part of the answer lies in the repetition across objects. Julia calls Allison Smith's recreation of handmade gas masks in her *Needle Work* project 'queer', implicitly referencing a non-nostalgic relation to the absent original. Being 'lost' in fact names the act of removal that the exhibition performs. All of these objects are lost, all out of context, all take up contrary relations to originality; all signify as prosthetics, as parts not wholes, as fragments of a broken vessel that cannot and must not be fixed. The brokenness of protest is part of what we celebrate here, the failure that registers as resistance to the whole notion of winning in the first place. And what objects are missing from this show because they are not obviously and assertively 'political'? What about dildos, drag queen costumes, burned bras, punk safety pins, 'zines, smashed guitars and torn T-shirts? What are the lost objects of this exhibition? What is not here because something else is?

In order to make worlds, other worlds must be unmade; new memories require forgetting; new paths require us to get lost. When objects are disobedient they also perform some of this unmaking, and so, as much as the objects assemble, produce, create, gesture, represent and speak, they also collapse, fall silent, sit still, shatter, obliterate, randomize and disappear. They represent not only the presence and here-ness of protest, but also the absence and the disintegration of dreams, fantasies and aspirations. Objects must certainly affirm our protestations with an affirmative and joyous 'Yes!'; they should certainly register our despair with a resounding 'No!'; they must contain our ambivalence with a wavering 'Maybe'. But objects, being objects, collectively perform acts of silence and stillness that living humans can only ever approximate.

Julia Bryan-Wilson

I have a quick observation at the level of a meta-comment that spins into a series of questions, so bear with me ... This discussion has attended to both the titular terms of the exhibition, as we have all tried to tackle our multiple understandings of 'disobedience' and 'objects' (their capacities to speak, as well as their stubborn muteness – thanks for that, Jack). But it leads me to reflect upon some of the absent, haunting terms here, especially the spectre of 'obedient objects'. What might those look like? Are they not in some cases the same as disobedient objects?

It is crucial to consider the realms of ideology, enforcement, policing, law and regulation that might compel or enforce obedience unequally, putting extra pressure on certain subjects. What sorts of privileges sometimes accompany acts of disobedience? How is obedience sometimes a necessary form of survival?

Finally, do we assume that only subjects and objects can object (to use the verb form)? And how does one account for the disobedient intangibles, like disobedient sounds, or disobedient moods?

Jack Halberstam

Julia, the question of 'obedient objects' is a really good one, as are the other questions that you pose about the necessity of certain forms of obedience for heavily policed subjects. I guess the category of 'obedient objects' functions like all kinds of normative classifications that are only glimpsed as constructions because we see that they depend upon certain constitutive exclusions. 'Obedient objects', in other words, could be everything ... and nothing. As you say, some of the disobedient objects – especially the craft ones, like quilts and embroidery, teapots and jewellery – could certainly function as obedient as well. But nothing about their creation necessarily speaks and signifies obedience in the way that something might signal refusal within the category of the 'disobedient'.

So, what do you imagine would round out this category, especially for the purposes of display in a museum? Torture instruments, canes, classroom implements, prison keys? Since these kinds of objects are not constructed under the same kind of pressure and with the same fierce aesthetic and political commitments that dissidents and prisoners, protesters and radicals bring to the creation of their 'disobedient objects', would they even form a genre at all? Is there anything that would bind 'obedient objects' to one another, or to obedience for that matter?

The mood/sound questions are awesome, too, but I will leave those to others and end with my own question: how do these objects reimagine or help us to reimagine not just politics but the act of protest itself? As T.V. stated, modes and methods of protest, as well as their slogans and chants, have quickly become stale and redundant. What styles, modes and aesthetics of protest might this collection summon? What is the role of aesthetics in creating an effective challenge? Can we think differently about protest and collectivity?

Julia Bryan-Wilson

What an interesting response. Actually, I was not thinking of 'obedient objects' as the literal, material culture of discipline – i.e. leg braces or prison keys – but, rather, as what we generally speak of as 'art', in its most normative (if perpetually contested) definition. Paintings, sculptures and other works created explicitly in relation to the realm of the aesthetic are often taken as quiescent, content to hang on living room walls or sit in galleries, in contradistinction to, say, scrawled signs used for street demonstrations. Of course, the art history I am invested in argues vigorously against this assumption, but it might be useful to confront, and more fully flesh out, the status of 'art' and, as Jack suggests, of aesthetics in this exhibition and in resistant, 'disobedient' cultures. As artists like Glenn Ligon are aware, in a painting like *Untitled (I Am A Man)*, from 1988, protest signs have their own dense visual appeal. The graphic look of such protests is crucial to their power.

T.V. Reed

We have rightly been raising questions about what the context of an aesthetic site, and a historically rather conservative one at that, might do to our 'disobedient objects', but what of the reverse? We are dealing with an ongoing process, one in which seriously critical works of 'art' are involved just as much as our putatively non-aesthetic objects of protest. 'Art' objects and protest signs may not be that far apart. Art objects that were once aimed at the heart of capitalism now adorn the walls of multinational corporations. All one has to do is trace the fate of an artist like Picasso – a revolutionary artist who was also at times an artist revolutionary – to understand that there is no resolution to this situation, but only an ongoing dialectic (often in recent decades at least a 'negative dialectic', as Theodor Adorno articulated with such richly obscure clarity). A negative dialectic perhaps includes Jack's 'Maybe' and many kinds of silence. We need both positive and negative forms of creativity to make what a wonderful recent handbook of protest calls *Beautiful Trouble*.[3]

Take an object like *Guernica*. A brilliant piece of protest/art that has many times moved from Adorno to adorn and back again. Even the most highly resistant art object can be captured by aesthetic neutralization, or museumization. But the art objects and the protest signs decidedly speak back to this process. Picasso was in Nazi-occupied Paris during much of World War II, and was, because of his 'decadent' art and association with 'known Jews', subject to periodic harassment by the SS. During one search of his apartment, an officer saw a photograph of *Guernica*. 'Did you do that?' the German asked Picasso. 'No,' he replied, 'You did.' Yes indeed.

Until 2009 a tapestry reproduction of *Guernica* hung in the UN headquarters in New York, and recall that the stage managers of the Empire during the dark Bush years understood, as had the Nazis, that the artwork had power. This copy of *Guernica* was sitting rather tamely in the UN building becoming 'classical', when in February 2003 the US government feared its power enough to cover it up when one of its agents delivered a new batch of lies to help launch hideous new war crimes in Iraq. But like much censorship before and since, this only served to reawaken resistance in the object and its advocates. Outside the building, a group of protesters held up copies of the painting, and soon a widely circulated protest poster, *Iraqnica* (credited to Russell Donegon and Plastic Jesus) drew precisely the link the Bush league sought to cover up. The image went viral and boosted an emergent anti-war movement. So let us not 'misunderestimate' the power our objects may have, even on Victoria and Albert. For both 'art' and 'disobedient objects' of protest, the process of incorporation and reinvention is an ongoing one that with his typically preternatural concision Walter Benjamin summarized in a one-liner: 'When politics becomes aestheticized, art must become politicized.' Art is very much part of the struggle, maybe even the same struggle, as our presumptively non-aesthetic objects of disobedience. Julia is quite right that effective protest needs artfulness, today more than ever when protests can be brushed aside by the lamestream media (the one concept I [sort of] share with Sarah Palin) as 'sixties style demonstrations', as if protest were just a retro style choice. When the students in Paris in 1968 shouted 'All power to the imagination!' they understood that the inability to imagine alternatives to current social formations is at the heart of oppression, and 'art' along with and often as part of protest movements is therefore at the heart of successful resistance and the building of new worlds. If 'another world is possible', and it is, it begins in the imagination grounded in real worlds of pain, beauty and fiercely practical hope.

John Holloway

Disobedient words are what we want. Disobedient sounds, yes, if we could sing or speak or scream as a sonic accompaniment. But it is written words that we are invited to produce for the exhibition, words to be published in a book that will go on sale (or perhaps be stolen). How can we write words worthy of the objects that are on display? A fearsome, exhilarating challenge.

How to write words that will be part of a museum publication but at the same time leap off the page and shout 'do not close the covers of this book, do not put it away on the shelf, do not put it on the coffee table. Do not forget what you have seen, carry it with you into your daily joys and angers'?

I picture us as part of a chorus. The lead singers are the objects on display and we stand in the background, talking to each other of course, as we are now, but aware too that we are talking to those who visit the exhibition, trying to find ways of amplifying the message of the objects.

In some cases, we can only stay silent. When Gavin tells us the story of Herman Wallace's jewellery, what else can we do but fall silent and scream to high heaven?

Each object speaks of its own particular dignity. But the power of the exhibition is surely that, in bringing the objects together, it de-objectifies them, reactivates them as Julia put it, brings them back to the diverse but collective subjectivity of those who made them and carried them, a rebel subjectivity of which we and the visitors are a part. As we see one object after another, we see a multiplicity of struggles, but not just that. Each object looks at the others and says, 'I am not alone, you are not alone, we are not alone.' And we, invited to participate from afar, look in our mind's eye at the objects, and the struggles they portray, and we say, 'I am not alone, you are not alone, we are not alone.' And the visitors, seeing the objects and reading these words, cry, 'I am not alone, you are not alone, we are not alone.' And then we all laugh for joy, so loud that capitalism falls apart and we make the world our own. That is what will happen if the exhibition succeeds. That (in response to Gavin and Turbulence) is what it would mean to win.

As part of the process, let Julia's wish come true: 'Let's hope that the Museum itself becomes disobedient [we might want to re-name it on the way – Emma & Karl, perhaps?], fostering new kinds of conversations around these objects and performances and histories, and reaching out in radical ways to new publics.' Then, indeed, our disobedient words would leap off the page and join hands with the disobedient objects escaped from their cases to dance a new world into existence.

AFTERWORD

Disobedient Objects gives a history of design told from below, through the objects produced by grassroots social movements. These are often simple in means, but rich in purpose and in impact. Powerful, provocative, poignant or subversively humorous, these objects challenge the ways in which we look at design – and at each other.

The majority of the objects in the exhibition have been loaned directly by activist groups. They raise difficult questions about the world we live in. Several were made during recent struggles that are unresolved or still raw. In many cases they encapsulate the jeopardy and trauma experienced by their makers. We are immensely grateful to these groups and individuals for sharing their work and ideas with the Victoria and Albert Museum. This is a brave and unusual exhibition showing work by brave and unusual designers. We are proud to present their work to the public.

The V&A is a historic institution with a radical mission: to bring art and design to all. Prince Albert, the museum's founder, was inspired by the work of Gottfried Semper, who described museum collections as 'the true teachers of a free people'. Today, we stay true to that mission by recognizing ongoing struggles for freedom taking place around the world.

Disobedient Objects reveals design to be much more than just a professional practice or a commercial process. It shows that even with the most limited of resources, ordinary people can take design into their own hands. It celebrates the creative 'disobedience' of designers and makers who question the rules.

Martin Roth
Director, Victoria and Albert Museum

ACKNOWLEDGEMENTS

We would like to thank all the groups and individuals who lent objects to the exhibition for their trust and good faith, and all of those who participated in our PAR workshops for their time, energy and care.

Glenn Adamson, Guy Atkins, Roberta Bacic, Jonathan Barnbrook, Tony Bennett, Gillian Boal, Iain Boal, Leah Borromeo, Andrew Boyd, Gavin Brown, Thalia Campbell, Ivan Cash, Chris Chalkley, Chickpea and the Dissident Island crew, Danielle Child, Seijin Cho, Joel Colover, Jeff Conant, Caroline Coon, Luke Cordingley, Alberto Corsin, Margaret Cox, Steve Crawshaw, B Dahl, Andy Dao, Hannah Davey, Jeremy Deller, Bill DiPaola, Isabeau Doucet, Marina Drummer, Tim Edkins, Adolfo Estalella, Ibrahim Fakhri, Ann Field, Grey Filastine, Phil Gard, Tim Gee, Tilly Gifford, Eldan Goldberg, Malu Halasa, Ed Hall, Tríona White Hamilton, Greta Hansen, Stefano Harney, Penelope Harvey, Sharon Hayashi, Jenny Holdcroft, Magdalena Jitrik, Jason Jones, John Jordan, Guy Julier, George Katsiaficas, Yaman Kayabali, Narita Keisuke, Dorothy Kidd, Dongwon Kim, Robert Hilary King, Hannah Knox, Rob Logan, Leónidas Martin, Cat Mazza, Josh McPhee, Andre Mesquita, Mika Minio-Paluello, Nick Mirzoeff, Helen Molesworth, Dave Morris, Zaher Omareen, Annabelle Parker, Richard Pell, Paul Polyp, Theo Price, Julia Ramirez, Mark Read, Carrie Reichardt, LJ Roberts, Gregory Sholette, Shehnaz Suterwalla, Nato Thompson, Helen Yaffe, Donna Yates.

V&A

We would like to thank the many colleagues at the V&A who have helped us to realize this project. In particular Dana Andrew, Ruth Cribb, Natalia Ferreiro, Ann Hayhoe, Amy Higgitt, Lizzie Hines and Sarah Jameson in the Exhibitions Department, Kieran Long, Corinna Gardner and Louise Shannon in Contemporary Architecture Design and Digital, Clair Battisson in Conservation and Richard Davis and George Eksts for photography.

Our special thanks go to our Research Assistant Steffi Duarte and editor Philip Contos. We are extremely grateful to Marwan Kaabour at Barnbrook for the design of this book.

Contributors

Julia Bryan-Wilson is associate professor of modern and contemporary art at University of California, Berkeley. She is the author of *Art Workers: Radical Practice in the Vietnam War Era* (2009), and editor of *OCTOBER Files: Robert Morris* (2013). She has published widely on feminist issues and is currently working on a book about contemporary textiles.

Anna Feigenbaum is a writer, researcher and educator based at Bournemouth University. She is co-author of the book *Protest Camps* (2013) and is currently writing a social history of tear gas (2015).

Catherine Flood is a Prints Curator in the Word and Image Department at the V&A and author of *British Posters: Advertising, Art & Activism* (V&A, 2012). She has curated exhibitions and displays at the V&A on subjects including propaganda, Victorian sentimentality and satire in fashion. In 2009, she coordinated Designing Democracy, a research project on posters produced during the collapse of communism in Europe.

David Graeber is an anthropologist and activist currently teaching at the London School of Economics. He is the author of a number of books, including *Lost People* (2007), *Possibilities* (2007), and *Debt: The First Five Thousand Years* (2011), and has participated, sometimes more, sometimes less effectively, in a number of social movements from the Direct Action Network to Occupy Wall Street. He is currently working on books about bureaucracy, bullshit jobs and the origins of social inequality.

Gavin Grindon is Visiting Research Fellow at the V&A and Postdoctoral Fellow in Visual and Material Culture at Kingston University. He has published in the *Oxford Art Journal*, *Third Text*, *Art History and Radical Philosophy*; co-authored *A User's Guide to ~~Demanding~~ the Impossible* (2011); and is currently preparing a monograph on *Art and Activism*.

Jack Halberstam is Professor of American Studies and Ethnicity, Gender Studies and Comparative Literature at the University of Southern California. Halberstam is the author of five books: *Gothic Horror and the Technology of Monsters* (1995), *Female Masculinity* (1998), *In a Queer Time and Place* (2005), *The Queer Art of Failure* (2011) and *Gaga Feminism: Sex, Gender, and the End of Normal* (2012).

John Holloway is a Professor of Sociology at the Instituto de Ciencias Sociales y Humanidades in the Benemérita Universidad Autónoma de Puebla, Mexico. He has published widely on Marxist theory, on the Zapatista movement and on the new forms of anti-capitalist struggle. His books *Change the World without Taking Power* (2002, new edition 2010) and *Crack Capitalism* (2010) have stirred an international debate.

Ana Longoni is a writer, researcher of CONICET and professor at the Universidad de Buenos Aires. She also teaches at the Programa de Estudios Independientes (Independent Studies Programme) of MACBA (Barcelona) and other universities. Her special field of research is the crossroads between art and politics in Latin America.

Francesco Raparelli has long been active in student and labour protest movements, has directed the *Centro studi per l'Alternativa comune*, and now takes part in the research activities of the Swiss Institute in Rome. He has published *La lunghezza dell'Onda: Fine della sinistra e nuovi movimenti* (2009) and *Rivolta o barbarie: La democrazia del 99 per cento contro i signori della moneta* (2013).

T.V. Reed is Buchanan Professor of American Studies and English at Washington State University. He is a lifelong activist-scholar whose books include *Fifteen Jugglers, Five Believers: Literary Politics and the Poetics of American Social Movements* (1992) and *The Art of Protest: Culture and Activism from the Civil Rights Movement to the Streets of Seattle* (2005). His most recent book is *Digitized Lives* (2014).

Nicholas Thoburn has published widely on media aesthetics, the politics of objects and extra-parliamentary movements. He is the author of *Deleuze, Marx and Politics* (2003), co-editor of *Objects and Materials* (2014) and is a Senior Lecturer in sociology at the University of Manchester. He is currently writing a monograph on political publishing.

Mark Traugott is Professor of History and Sociology at the University of California at Santa Cruz. His 2010 book, *The Insurgent Barricade*, examines how a European culture of revolution was established and sustained and what the recurrence and persistence of barricade-building can teach us about how social contention originates and spreads.

Introduction

1 Peter Linebaugh and Marcus Rediker, *The Many-Headed Hydra: Sailors, Slaves, Commoners, and the Hidden History of the Revolutionary Atlantic* (Boston, 2001), pp.3–4.

2 Though the story of the Museum (and of the culture of rebellions) is not so clear-cut as this opposition might suggest. Beyond libertarian socialist William Morris's well-known contributions to the Museum, one of its founding figures, the architect Gottfried Semper, came to work in England fleeing in exile from Germany, after he took part in the failed 1849 May uprising in Dresden alongside Mikhail Bakunin. During the revolt he turned his architectural intellect to the construction of barricades, including one grand effort that became known as the famous 'Semper Barricade', which he personally defended for some days. Later, from 1922, Herbert Read's employment as keeper in the Museum also afforded him the space to develop his key ideas and writing on art and anarchism, first in pamphlets for Freedom Press and later in the collection *To Hell With Culture*, which anticipated many central ideas on culture and creativity upheld by the 'new anarchism' of the 1960s onwards. From 1947 the V&A Circulation Department featured several Communist Party members and, led by active Communist and Morris expert Peter Floud, hired women with art-school backgrounds instead of Oxbridge 'gentleman keepers', and focused on modern objects that contested the prevailing tastes and hierarchies of the Museum.

3 'Social movement' is a sociological term for the organizations behind what is commonly called 'protest' or 'direct action'. Sociologists still debate its exact definition and bounds, but essentially it encapsulates large informal and non-institutional groups of people concentrating on political and social issues.

4 Tim Gee, *Counterpower: Making Change Happen* (London, 2011).

5 Archives and collections such as the International Institute for Social History, Amsterdam; Interference Archive, New York; People's History Museum, Manchester, and Mayday Rooms, London, are to be celebrated for their documentation and preservation of the material of movement cultures, and this project stands on the shoulders of such work.

6 Josh McPhee and Dara Greenwald, *Signs of Change: Social Movement Cultures 1960s to Now* (Edinburgh, 2010). On this stereotyping, see Imogen Tyler, *Revolting Subjects: Social Abjection and Resistance in Neoliberal Britain* (London, 2013).

7 *The Interventionists: Art in the Social Sphere*, Massachusetts Museum of Contemporary Art, 2004–5; *Forget Fear*, the 7th Berlin Biennale, KW Institute, Berlin, 2012.

8 Examples can be found in *Design for the Other 90%* (New York, 2007); *Design for the Elastic Mind* (New York, 2008); Ann Thorpe, *Architecture and Design versus Consumerism: How Design Activism Confronts Growth* (London, 2012).

9 See, in this regard, Andrew X, 'Give Up Activism', *Do or Die* 9, pp.160–6.

10 That is, working from the perspective De Angelis calls 'the beginning of history'. Massimo De Angelis, *The Beginning of History: Value Struggles and Global Capital* (London, 2006).

11 David Harvey, *A Brief History of Neoliberalism* (Oxford, 2005).

12 Naomi Klein, *The Shock Doctrine* (London, 2008).

13 Peter Kornbluh, *The Pinochet File: A Declassified Dossier on Atrocity and Accountability* (New York, 2013).

14 It is in such a scream of refusal and hope that John Holloway theorizes the root of disobedient social agency. John Holloway, *Change the World Without Taking Power* (London, 2002).

15 By 'aesthetic composition' we intend something close to Raymond Williams's 'structure or feeling', but emphasizing affect and aesthetics not only as a determined structure (whether emergent or otherwise) but also as bound to agency and processes of political/class composition. Raymond Williams, *The Country and the City* (Nottingham, 2011). See also Deborah Gould, *Moving Politics* (Chicago, 2009).

16 Judith Jack Halberstam, *The Queer Art of Failure* (Durham NC, 2011).

17 Stephen Duncombe and Steve Lambert, 'An Open Letter to Critics Writing About Political Art'; artisticactivism.org/2012/10/an-open-letter-to-critics-writing-about-political-art/ (accessed 10 December 2013).

18 Authors' conversation with the artist, October 2013.

19 Rebecca Solnit, *Hope in the Dark: The Untold History of People Power* (Edinburgh, 2005).

20 David Graeber, *Ask Me Anything*; libcom.org/forums/general/david-graebar-ama-28012013 (accessed 10 December 2013).

21 I am grateful to Carwil Bjork-James for his expertise on the role of Wiphalas in social movements; woborders.wordpress.com/ (accessed 10 December 2013).

22 Ayatollah Ahmad Jannati called for the shoes, 'more valuable than crowns', to be placed in an Iraqi museum, but they were destroyed by United States security forces.

23 Likewise, in Latin America, a traditional tool by which farmers earned their income, embedded in various local popular traditions, has also long held a powerful symbolic role in protests: the machete. In 1959 Fidel Castro announced the machete as the symbol of the Cuban revolution, echoing the use of ploughs, sickles and hammers in European and Russian socialist iconography. The machete's symbolic function is supported by its long history as a weapon for poor farmers and freed slaves in wars of independence. Still in practical use, machetes are often carried symbolically in demonstrations rather than depicted graphically. By 2012, they could be seen in student demonstrations in Mexico paired with Guy Fawkes masks.

24 For an introduction to the idea of a 'circulation of struggles', see Nick Dyer Witherford, *Cyber-Marx: Cycles and Circuits of Struggle in High-Technology Capitalism* (Chicago, 1999).

25 Although this design spread, it was pre-dated or paralleled by similar architectural design solutions to resisting eviction, for example, the nets and scaffold towers of the anti-road building actions of Claremont Road, London, in 1994; the Sharpness nuclear train blockade of 1980 in Britain; or the wooden towers and barriers of the anti-nuclear Free Republic of Wendland, Germany, in 1980. Ian Cohen, *Green Fire* (Pymble, 1997), p.199; Iain McIntyre (ed.), *How to Make Trouble and Influence People* (Oakland, 2013) p.154.

26 Rob Evans and Paul Lewis, *Undercover: The True Story of Britain's Secret Police* (London, 2013).

27 See David Graeber's essay in this volume (p.68).

28 'Armed Police on May Day Riot Alert', *Observer*, 22 April 2001.

29 Bonnie J. Dow, 'Feminism, Miss America, and Media Mythology', *Rhetoric & Public Affairs* (2003), 6.1: 127–49.

30 'One of the paternalistic ideas I have often heard from academics is that cutting edge political thinking takes place in the academy. I have found the opposite to be true – that it takes place outside of the academy where it is not hampered by institutional requirements, such as the focus on individual scholarship [or] the need to develop special vocabularies and grand theory in order to be taken seriously.' Maxine Wolfe, 'Inside/Outside the Academy: The Politics of Knowledge in Queer Communities'; www.actupny.org/documents/academia.html (accessed December 2013).

31 Karl Marx, *Grundrisse: Outlines of the Critique of Political Economy*, trans. Martin Nicolaus (London, 1973), pp.690–712.

32 Stefano Harney and Fred Moten, *The Undercommons: Fugitive Planning and Black Study* (London, 2013), p.26.

33 See, for example, senior curator Kieran Long's first column for *Dezeen*; www.dezeen.com/2013/09/12/opinion-kieran-long-on-contemporary-museum-curation/ (accessed 10 December 2013).

34 Tony Bennett, *The Birth of the Museum* (London, 1995), p.21.

35 Gerald Raunig, 'Instituent Practices: Fleeing, Instituting, Transforming', *Art and Contemporary Critical Practice: Reinventing Institutional Critique*, ed. Gene Ray and Gerald Raunig (London, 2005).

36 Michael Hardt, *The Procedures of Love*, dOCUMENTA (13): 100 Notizen – 100 Gedanken [100 Notes, 100 Thoughts] (Kassel, 2012); Jamie Heckert, 'Listening, Caring, Becoming: Anarchism as an Ethics of Direct Relationships', in Benjamin Franks and Matthew Wilson (eds), *Anarchism and Moral Philosophy* (London, 2010).

37 Douglas Crimp, *On the Museum's Ruins* (Cambridge MA, 1993).

38 One other non-public site which we did not pursue was the undisclosed holdings of movement objects seized or stolen by the state, though we found traces of these ghost archives in stories of an undocumented confiscated protest banner used as mocking decoration in a British police station; the Chinese government's alleged longstanding archive of objects handed in or left on the street after demonstrations; or the Occupy Wall Street barriers that reappeared outside the police chief's office on their anniversary, as recorded in 'Whodunit at Police Headquarters: Occupy Accuses Police', *New York Times*, 17 November 2012.

39 Gregory Sholette, 'Not Cool Enough to Catalog' and John Jordan, 'On Refusing to Pretend to do Politics in a Museum', *Art Monthly* (March, 2010), no.334.

40 Brian Holmes, *'Liar's Poker', Unleashing the Collective Phantoms* (New York, 2007).

41 Pierre-Olivier Rollin, 'The Return of Temples of Culture: from Museum to Mausoleum', in Lieven De Cauter, Ruben De Roo and Karel Vanhaesebrouck (eds), *Art and Activism in the Age of Globalization* (Rotterdam, 2011), pp.74–82.

42 Ibid.

Barricades as Material and Social Constructions

1 These subjects are treated at greater length in Mark Traugott, *The Insurgent Barricade* (Berkeley CA, 2010).

2 The 1694 *Dictionnaire de l'académie française* explicitly defined a barricade in this way: 'Type of entrenchment that is usually made with barrels filled with earth for the purpose of defending oneself or finding shelter from the enemy.' Académie française, *Dictionnaire de l'Académie française* (Paris, 1694).

3 It seems superfluous to cite evidence of the presence of traditional barricades in 1830 or 1848, since period illustrations as well as eyewitness accounts are so plentiful and readily available. The illustration on p.28 is perhaps sufficient to document their use in 1648. Since there are no contemporaneous images of the 1588 uprising – the image on p.27 actually dates from the nineteenth century – the reader may wish to refer to such sources as Pierre Victor Palma Cayet, 'Introduction' to *Chronologie novénaire* (1608), in J.-F. Michaud and J.-J.-F. Poujoulat (eds), *Nouvelle collection des memoires pour servir à l'histoire de France, depuis le XIIIe siècle jusqu'à la fin du XVIIIe siècle*, 12:1 (Paris, 1838), pp.13–86; Pierre de L'Estoile, *Journal de L'Estoile pour le regne de Henri III, 1574–1589* (Paris, 1943), p.554; and Nicolas Poulain, 'Le Procez-verbal d'un nommé Nicolas Poulain, Lieutenant de la Prévoste de l'Isle-de-France, qui contient l'histoire de la Ligue, depuis le second janvier 1585 jusques au jour des barricades, escheues le 12 may 1588' (1588) in M.L. Cimber and F. Danjou (eds), *Archives curieuses de l'histoire de France depuis Louis XI jusqu'à Louis XVIII* (Paris, 1836), pp.299–300. As for the barricades of 1789, which historians have almost universally overlooked (when they have not denied their existence outright), one can find ample testimony in Charles-Louis Chassin (ed.), *Les elections et les cahiers de Paris en 1789* (Paris, 1888–9), vol.3, pp.539–40; Jacques Godechot, *The Taking of the Bastille, July 14th, 1789* (New York, 1970), p.204; Louis-Philippe, King of the French, *Memoirs, 1773–1793*, translated by John Hardman (New York, 1977), p.44; and Paris, Assemblée générale des électeurs, *Procès-verbal des séances et délibérations de l'Assemblée générale des électeurs de Paris, réunis à l'Hôtel de Ville le 14 juillet 1789* (Paris, 1790), p.276, among other works.

4 For more details on barricade components, see Traugott, cited note 1, pp.14–15, as well as the many entries concerning barricade materials to be found in the index of that book.

5 Traugott, cited note 1, pp.7–10.

6 I have identified 155 barricade events that occurred in Europe through the end of the nineteenth century, 92 of them in France. Of these, no fewer than 23 occurred in Paris.

7 See Charles Jeanne, *A cinq heures nous serons tous morts! Sur la barricade Saint-Merry, 5–6 juin 1832*, ed. Thomas Bouchet (Paris, 2011).

8 Compare, for example, Maxime Du Camp, *Souvenirs de l'année 1848: La revolution de février, le 15 mai, l'insurrection de juin* (Paris, 1876), pp.265–6.

9 Auguste Blanqui, *Instructions pour une prise d'armes; L'éternité par les astres: Hypothèse astronomique, et autres texts*, eds Miguel Abensour and Valentin Pelosse (Paris, 1973).

10 A highly useful discussion of these and related tactical considerations can be found in the introduction to the forthcoming book by Jonathan M. House, *Controlling Paris: Armed Forces and Counter-Revolution, 1789–1848* (New York, 2014).

11 The supposed advantages of these contrivances, also referred to as 'counter-barricades', were discussed in *Journées illustrées de la révolution de 1848: Récit historique de tous les événements ... accompagné de 600 gravures, L'Illustration*, Paris (1848–9), pp.263–4.

12 Though Napoleon's tactical improvisation has become notorious, most commentators seem unaware that the royalist insurgents he was seeking to repress were in the act of constructing a barricade. See Paul Barras, *Rapport fait à la Convention nationale le 30 Vendémiaire, an 4 sur la conspiration et la rébellion qui ont éclaté dans les journées des 13 et 14 Vendémiaire, et sur les opérations militaires executées par l'armée républicaine* (Paris, 1795), p.12.

13 See, merely by way of example, these 2012 articles on the use of burning-tyre barricades in protests against austerity in Spain and in demonstrations against increases in bus fares in Mozambique: rt.com/news/spanish-miners-clashes-cuts-646/; and www.reuters.com/article/2012/11/15/us-mozambique-protests-idUSBRE8AE0SM20121115. With regard to the Reclaim the Streets movement in London and New York, I am grateful to Gavin Grindon for having pointed out the parallels between its use of 'human tripods' and classic barricades. See, for example, www.fraw.org.uk/files/direct_action/tripod.html; beautifultrouble.org/case/reclaim-the-streets/; www.culturechange.org/issue10/street-parties.htm; and times-up.org/ongoing-demonstrations/reclaim-streets.

14 For comparative evidence in support of this proposition, see Katherine Chorley, *Armies and the Art of Revolution* (Boston, 1973).

15 I have in mind not just weapons in the strict sense, but also improved means of transport and communication. In the nineteenth-century context, the most important of these changes resulted from the introduction of the railroad and the telegraph. While in theory these might have yielded benefits for either side in a violent conflict, in actual practice, the armed forces were in a stronger position to exploit the enhanced logistical capacity that they made possible.

16 See, for example, Thomas-Robert Bugeaud, *La guerre des rues et des maisons* (Paris, 1997). An insightful discussion of these kinds of military advances in the context of the defence of nineteenth-century Paris can be found in the previously cited work by House (note 10).

The Disobedient Objects of Protest Camps

1 Anna Feigenbaum, Fabian Frenzel and Patrick McCurdy. *Protest Camps* (London, 2013).

2 J. Juris, *Networking Futures: The Movements Against Corporate Globalization*, (Durham NC, 2008); B. Hooks, *Yearning: Race, Gender, and Cultural Politics* (Boston, 1990).

3 Zoe Sofia, 'Container Technologies', *Hypatia* (2000), 15: 181–201.

4 Ian Hodder, *Entangled: An Archaeology of the Relationships between Humans and Things* (Malden MA, 2012).

5 Ruth Cowan, *More Work For Mother: The Ironies of Household Technology from the Open Hearth to the Microwave* (New York, 1983).

6 Matthew Collin, *The Time of the Rebels* (London, 2007).

7 Road Alert!, *Road Raging: Top Tips For Wrecking Road Building* (1997); www.eco-action.org/rr/ (accessed 24 May 2013).

8 Fabian Frenzel, 'Politics in Motion,' PhD Thesis, Leeds Metropolitan University Library, 2010.

9 Gregory Cowan, 'Nomadology in Architecture: Ephemerality, Movement and Collaboration', MSc dissertation, University of Adelaide, 2002.

10 J. Filip, 'Judge Lets Occupy Fort Myers Back into Park', *Court House News*, 2011; www.courthousenews.com/2011/11/23/41704.htm (accessed 24 May 2013).

11 Caroline Davies, 'Occupy London Protesters to Appeal Against Eviction Ruling', *Guardian*, 18 January 2012; www.guardian.co.uk/uk/2012/jan/18/occupy-london-protesters-appeal-eviction (accessed 24 May 2013).

12 Feigenbaum, Frenzel, McCurdy 2013, cited note 1.

13 Protest Camps.org, 'Spontaneous Urban Design: An Interview with 123 Occupy', protestcamps.org/2012/05/01/spontaneous-urban-design-an-interview-with-123-occupy-2-2/ (accessed 24 May 2013).

14 Barbara Ehrenreich, 'Throw Them Out with the Trash: Why Homelessness is Becoming an Occupy Wall Street Issue', tomdispatch.com, 23 October 2011; www.tomdispatch.com/blog/175457/ (accessed 24 May 2013).

15 Diana Denham and C.A.S.A. Collective, *Teaching Rebellion: Stories from the Grassroots Mobilization in Oaxaca* (Oakland, 2008).

16 R. Nunes, 'The Intercontinental Youth Camp as the Unthought of the World Social Forum,' *ephemera: theory & politics in organization* 5/2 (2005), 277–296.

17 'The Dongas Tribe in the Summer of 1996', www.youtube.com watch?v=CwVexEZ8K8c (accessed 24 May 2013).

18 Isy Morgenmuffel, 'The Feeding of the Five Thousand' in David Harvie et al (eds), *Shut Them Down! The G8, Gleneagles 2005 and the Movement of Movements* (Leeds and New York, 2005), p.173.

19 Anna Feigenbaum, Fabian Frenzel and Patrick McCurdy, 'Towards a Method for Studying Affect in (micro)Politics: The Campfire Chats Project and the Occupy Movement', *Parallax*, 19/2 (2013), 21–37.

20 N. Couldry, 'Disrupting the Media Frame at Greenham Common: A New Chapter in the History of Mediations?' *Media, Culture & Society* 21/3 (1999), 337–358, p.339.

21 *Morgenmuffel*, no. 14, January 2006.

22 Yvonne Marshall, *Archaeologies of Resistance* (Southampton, 2006), pp.8–9.

From *Tute Bianche* to the Book Bloc

1 This term, which became common, was introduced during the 2001 G8 Summit Meeting in Genoa, Italy, following the successes of the Seattle WTO protests. It described the area around the G8 meeting site which was off-limits for non-residents and surrounded by a barricade, blocking all contact and communication between protesters and summit attendees.

2 Andrea Fumagalli, Maurizio Lazzarato, *Tute Bianche: Disoccupazione di massa e reddito di cittadinanza* (Rome, 2002).

3 www.wumingfoundation.com/english/wumingblog/?p=1014

4 Luca Casarini, *Genova dentro* (Rome, 2011).

5 Antonio Negri et al., *Controimpero* (Rome, 2002).

6 See Anja Kanngieser, *Experimental Politics and the Making of Worlds* (London, 2012).

7 Francesco Raparelli, *La lunghezza dell'Onda* (Milan, 2009).

8 Members of Wu Ming, the Italian collective of guerrilla novelists, were the first to use the term 'Book Bloc'.

9 www.youtube.com/watch?v=0pAea8o__b8

10 Michel Foucault, *Nietzsche, la généalogie, l'histoire* (Paris, 1971)

11 In London, a pamphlet on creative protest strategies distributed at occupied university art schools invoked the Italian Book Blocs and the Climate Camp shields. The Laboratory of Insurrectionary Imagination, *A User's Guide to ~~Demanding~~ the Impossible*, Minor Compositions/Autonomedia, London, 2011.

12 As one book documenting the phenomenon suggests. *Book Bloc: Le voci della protesta, da Omero a Wu Ming* (Lantana, 2013).

13 Francesco Raparelli, *Rivolta o barbarie* (Milan, 2012).

14 From 2011, pictorial shields (and shields per se) appeared to spread internationally, not necessarily using only book imagery. Among other examples: in November 2011, in post-Occupy Wall Street eviction demonstrations, shields appeared bearing slogans appropriated from Spanish Indignados protests – '*No vas a tener una casa en la puta vida*' ('You will never own a home in your whole fucking life'); in February 2012, Oakland's Book Blocs were replaced by coffin-shields in a demand to end police brutality; while in January 2012 in Brazil, the eviction of Pinheirinho favela was resisted by activists with shields made from halved plastic barrels.

On the Phenomenology of Giant Puppets

1 One effect of the peculiar definition of violence adopted by the American media is that Gandhian tactics do not, generally speaking, work in the United States. One of the aims of non-violent civil disobedience is to reveal the inherent violence of the state. Since the 1960s, however, the US media has simply refused to represent authorized police activity of any sort as violent. In the years immediately preceding Seattle, for instance, forest activists on the West Coast had developed lock-on techniques by which they immobilized their arms in concrete-reinforced tubing, making them at once obviously harmless and very difficult to remove. It was a classic Gandhian strategy. The police response was to develop what can only be described as torture techniques: rubbing pepper spray in the eyes of incapacitated activists. When even that didn't cause a media furore, many concluded that Gandhian tactics simply didn't work in America. It is significant that a large number of the Black Bloc anarchists in Seattle, who rejected the lock-on strategy and opted for more mobile and aggressive tactics, were precisely the forest activists who had been involved in tree-sits and lock-ons in the past.

2 Cited in *The Black Bloc Papers*, compiled by David and X of the Green Mountain Anarchist Collective (Baltimore, 2002), p.53.

3 I owe the phrase to Ilana Gershon.

4 Field notes extract, 31 July 2000.

5 Tony Blair, from 'Blair: Anarchists will not stop us', BBC News Online, 16 June, 2001; news.bbc.co.uk/hi/english/uk_politics/newsid_1392000/1392004.stm.

6 From 'Puppet Masters: Paper Hand Puppet Intervention brings its form of political theater back to Chapel Hill', *Independent Online*, 8 August 2001; indyweek.com/durham/2001-08-08/ae.html (accessed June 2004).

7 Similar themes recur in many interviews with radical puppeteers. This is from Mattyboy of the Spiral Q Puppet Theater in Philadelphia: 'OK, I'm 23. I've lost 13 friends to AIDS. This is wartime, it's a plague. This is the only way for me to deal with it. With puppets I create my own mythology. I bring them back as gods and goddesses' (cited in Daisy Freid, 'The Puppets are Coming', *Philadelphia City Paper*, 16–23 January 1997).

8 Videographers documented police commanders on the first day reassuring activists that the Seattle police 'had never attacked non-violent protesters and never would'. Within hours the same commanders had completely reversed course.

9 Blocking a street is in fact technically not even a crime in the United States, but an 'infraction' or 'violation': that is, the legal equivalent of jaywalking, or a parking ticket. If one violates such ordinances for non-political purposes, one can normally expect to receive some kind of ticket, but certainly not to be taken to a station or jailed.

10 Jose Martinez, 'Police Prep for Protests Over Biotech Conference at Hynes', *Boston Herald*, 4 March 2000.

11 *New York Times*, 6 June 2000, Corrections, p.A2. The original story, by Nicole Christian (4 June 2000, p.A6) was significantly entitled, 'Detroit Defends Get-Tough Stance'.

12 This document was transcribed and widely circulated on activist listserves at the time. According to one story in the *Miami Herald* (Joan Fleischman, 'Trade protesters mean business, analyst warns', 1 October 2003), it derived from 'retired DEA agent Tom Cash, 63, now senior managing director for Kroll Inc., an international security and business consulting firm'. Cash in turn claimed to derive his information from 'police intelligence' sources.

13 Brendan I. Koerner, 'Can Miami Really Ban Giant Puppets', *Slate*, 12 November 2003; www.slate.com/id/2091139.

14 There is also no clear evidence that 1960s protesters spat on soldiers any more than early feminists actually burned bras. The story seems to have emerged in the late 1970s or early 1980s, and, as the 2005 documentary *Sir! No Sir!* by Displaced Films nicely demonstrates, the only veteran who has publicly claimed this happened to him is likely to be lying.

15 One common fear is that wooden dowels used in their construction could be detached and used as cudgels, or to break windows.

16 Marc Cooper, 'Dum Da Dum-Dum', *Village Voice*, 16 April 1991, pp.28–33.

Unpopular Pamphlets

1 Guy Debord, letter to Ivan Chtcheglov, 30 April 1963; www.notbored.org/debord-30April1963.html (accessed 16 July 2013).

2 As a relative assessment, this is clearly not a hard rule. For a favourable consideration of magazine form, see Nicholas Thoburn, 'Ceci n'est pas un magazine: The Politics of Hybrid Media in *Mute* Magazine', *New Media and Society* 14/5 (2012), 815–31.

3 Cited in Christina Kiaer, *Imagine No Possessions: The Socialist Objects of Russian Constructivism* (London, 2005), p.1. The use of the racial trope of blackness here is disconcerting, for it appears to designate only an abject quality of the enslaved. That said, blackness figures later in the letter in a more affirmative manner, as the 'black work' of 'reprisal against [the commodity's] oppressors'. Aleksandr Rodchenko, *Experiments of the Future: Diaries, Essays, Letters, and Other Writings*, Alexander N. Lavrentiev (ed.), trans. Jamey Gambrell (New York, 2005).

4 Boris Arvatov, 'Everyday Life and the Culture of the Thing (Toward the Formulation of the Question)', trans. Christina Kiaer, *October* 81 (1997), 119–28, 121; Karl Marx, 'Economic and Philosophical Manuscripts', trans. Gregor Benton, in *Early Writings* (Harmondsworth, 1975), pp.352, 390.

5 Ibid.

6 Karl Marx, 'On the Jewish Question', in *Early Writings*, cited note 4, p.229.

7 Boris Arvatov, 'From *Art and Class* (1923)', in Stephen Bann (ed.), *The Tradition of Constructivism* (London, 1974), pp.45–6.

8 Ibid.

9 Fabian Tompsett, interview with author, 5 June 2007. I am most grateful to Fabian Tompsett for sharing with me his publishing experience and insight.

10 Karl Marx, letter to Wilhelm Blos, 10 November 1877, cited in Jacques Camatte, 'About the Organisation', in *Capital and Community: The Results of the Immediate Process of Production and the Economic Work of Marx*, trans. David Brown (London, 1988).

11 See Nicholas Thoburn, 'Do Not Be Afraid, Join Us, Come Back? On the "Idea of Communism" in Our Time', *Cultural Critique* 84 (2013), 1–34.

12 Unpopular Books' preface to Jean Barrot, *What is Situationism: Critique of the Situationist International* (London, 1987), p.2.

13 Jørgen Nash, cited in Mikkel Bolt Rasmussen and Jakob Jakobsen (eds), *Expect Anything Fear Nothing: The Situationist Movement in Scandinavia and Elsewhere* (Copenhagen, 2011), p.223. The Situationist practice of *détournement* – the appropriation and reuse of existing semiotic materials in new ensembles – is a step toward a 'literary communism' that intervenes in the formal conditions of language, wrenching it from the general exchangeability that communication shares with the commodity. Guy Debord and Gil J. Wolman, 'Methods of Détournement', in Ken Knab (ed. and trans.), *Situationist International Anthology* (Berkeley CA, 1981), p.11.

14 Asger Jorn, cited in Simon Crook, 'Moving Mountains: "Shamanic" Rock Art and the International of Experimental Artists', *Transgressions: A Journal of Urban Exploration* 4 (1998), pp.36–48, p.42. This split in the Situationist International organization over the revolutionary role of 'art' is addressed in Unpopular Books' first published book: Stewart Home, *The Assault on Culture: Utopian Currents from Lettrisme to Class War* (London, 1988).

15 Tompsett, interview with author, 5 June 2007.

16 See Asger Jorn, 'Critique of Economic Policy', trans. Fabian Tompsett, in *Transgressions: A Journal of Urban Exploration* 4 (1998), 13–35.

17 Asger Jorn, *Open Creation and its Enemies*, trans. Fabian Tompsett (London, 1994), p.47.

18 Joad Raymond, *Pamphlets and Pamphleteering in Early Modern Britain* (Cambridge, 2003).

19 Marshall McLuhan, *The Gutenberg Galaxy: The Making of Typographic Man* (Toronto, 1962), p.125.

20 Tompsett, interview with author, 5 June 2007. One cover places the pamphlet in relation to Jorn's wayfaring habits, with an image of a lithograph of Jorn astride a BSA motorcycle in front of the Eiffel Tower; the other indicates Jorn's interests in material form that I am pursuing here, with its image of his somewhat talismanic memorial stone for his syndicalist friend Christian Christensen.

21 Jacques Rancière, *The Nights of Labor: The Workers' Dream in the Nineteenth Century*, trans. John Drury (Philadelphia, 1989), pp.8, xxv.

22 Tompsett, interview with author, 5 June 2007.

23 Jorn, cited note 17, p.32.

24 Gilles Deleuze and Félix Guattari, *What Is Philosophy?*, trans. Hugh Tomlinson and Graham Burchill (London, 1994), p.167.

25 Tompsett, interview with author, 5 June 2007.

26 Peter Pels, 'The Spirit of Matter: On Fetish, Rarity, Fact and Fancy', in Patricia Spyer (ed.), *Border Fetishisms: Material Objects in Unstable Spaces* (New York, 1998), p.98.

27 Ibid., p.99, emphasis added.

28 Ibid., p.98.

29 Denis Hollier, 'The Use-Value of the Impossible', trans. Liesl Ollman, *October* 60 (1992), 3–24, p.22.

30 Ibid., p.23.

31 Infopool, 'Operation Reappropriation: Infopool @ Tate Modern, 9.2.2001'; www.infopool.org.uk/tate/htm (accessed 5 June 2008).

32 Ibid.

33 Arvatov, cited in Kiaer, note 3, p.68.

34 Infopool, cited note 31.

35 Tim Barringer, 'The South Kensington Museum and the Colonial Project', in Tim Barringer and Tom Flynn, *Colonialism and the Object: Empire, Material Culture and the Museum* (London, 1998), p.11.

Disobedient Bodies: Art Activisms in Argentina

1 Ana Longoni and Mariano Mestman, *Del Di Tella a Tucumán Arde* (Buenos Aires, 2000; revised ed., 2008).

2 Pilar Calveiro, *Poder y desaparición* (Buenos Aires, 1997).

3 Ana Longoni and Gustavo Bruzzone (eds), *El Siluetazo* (Buenos Aires, 2007).

4 *GAS-TAR* (*Grupo Artistas Socialistas-Taller de Arte Revolucionario*) and *CAPATACO* (*Colectivo de Arte Participativo-Tarifa Común*); both acronyms are puns.

5 *GAC*, *Pensamientos Prácticas Acciones* (Buenos Aires, 2009); www.archive.org/details/GacPensamientosPracticasYAcciones (accessed October 2013).

6 *HIJOS* is an acronym for [in English] Sons and Daughters for Identity and Justice Against Oblivion and Silence.

7 '*Escrachar*' is Argentine vernacular for 'exposing', 'uncovering'. The *escraches* arose from the need to obtain 'social condemnation' for the perpetrators of State terrorism who had been either pardoned or simply never brought to trial, thanks to the legal impunity granted them. The exposure protest disclosed the repressor's identity, his face, his address and, above all, his past as a repressor to his neighbours and co-workers, who knew nothing about his criminal record.

8 '*Multiplicidad*', paper published by Malba-Proyecto Venus, Buenos Aires, May 2003.

9 *GAC*, cited note 5.

10 '*El TPS*', 2004, available at tallerpopulardeserigrafia.blogspot.com.ar/2008/11/textos-sobre-el-tps.html, (accessed December 2013).

11 Ibid.

12 *Tucumán Arde*, the best-known collective action of the Argentinian avant-garde of the 1960s, sought to generate information to counter the propaganda of the Onganía dictatorship on the crisis affecting the population in Tucumán, a province in the north affected by the closure of many sugar factories.

13 Brian Holmes, 'Liar's Poker: Representation of Politics/Politics of Representation', *Springeren* 1 (2003).

14 A book, *Manual de Mapeo Colectivo* (Buenos Aires, 2013), describing their experiences with this cartographic technique has just appeared.

Roundtable

1 allisonsmithstudio.com/p-r-o-j-e-c-t-s/needle-work/view/75 (accessed 10/12/2013).

2 Turbulence (ed.), *What Would It Mean to Win?* (Oakland, 2010).

3 Andrew Boyd, ed., *Beautiful Trouble: A Toolkit for Revolution* (London, 2012).

- Anonymous, *The Occupation Cookbook: Or the Model of the Occupation of the Faculty of Humanities and Social Sciences in Zagreb* (London, 2011)
- Jess Bell and Dan Spalding, *Security Culture for Activists* (Oakland, n.d.)
- Andrew Boyd, ed., *Beautiful Trouble: A Toolbox for Revolution* (London, 2012)
- Kyle Bravo, *Making Stuff and Doing Things* (Portland, 2005)
- Julia Bryan-Wilson, *Art Workers: Radical Practice in the Vietnam War Era* (Berkeley, 2009)
- Carrot Workers Collective, *Surviving Internships: A Counter Guide to Free Labour in the Arts* (London, n.d.)
- Harry Cleaver, *Reading Capital Politically* (Edinburgh, 2000)
- Crimethinc Collective, *Recipes For Disaster: An Anarchist Cookbook* (Salem, 2012)
- Stephen Duncombe (ed.), *The Cultural Resistance Reader* (London, 2002)
- Anna Feigenbaum, Fabian Frenzel and Patrick McCurdy, *Protest Camps* (London, 2013)
- Tim Gee, *Counterpower: Making Change Happen* (London, 2011)
- Uri Gordon, *Anarchy Alive! Anti-Authoritarian Politics from Practice to Theory* (London, 2007)
- David Graeber, *Direct Action: An Ethnography* (Edinburgh, 2009)
- Dara Greenwald and Josh McPhee, *Signs of Change: Social Movement Cultures, 1960s to Now* (Edinburgh, 2010)
- Gavin Grindon (ed.), *Art, Production and Social Movement* (London, 2014)
- Judith Halberstam, *The Queer Art of Failure* (Durham, 2011)
- John Holloway, *Change the World Without Taking Power: The Meaning of Revolution Today* (London, 2002)
- Anja Kanngieser, *Experimental Politics and the Making of Worlds* (London, 2013)
- Kate Khatib, Margaret Killjoy and Mike McGuire (eds), *We Are Many: Reflections on Movement Strategy from Occupation to Liberation* (Edinburgh, 2012)
- The Laboratory of Insurrectionary Imagination, *A User's Guide to ~~Demanding~~ the Impossible* (London, 2011)
- Nicolas Lampert, *A People's Art History of the United States: 250 Years of Activist Art and Artists Working in Social Justice Movements* (New York, 2013)
- Alessio Lunghi and Seth Wheeler (eds), *Occupy Everything! Reflections on Why It's Kicking Off Everywhere* (London, 2011)
- Subcomandante Marcos, *Our Word is Our Weapon: Selected Writings* (London, 2002)
- Paul Mason, *Live Working or Die Fighting: How the Working Class Went Global* (London, 2008)
- Iain McIntyre, *How to Make Trouble and Influence People: Pranks, Hoaxes, Graffiti and Political Mischief Making from Across Australia* (Carlton, 2009)
- North East Forests Alliance, *The Intercontinental Deluxe Guide to Blockading* (updated); nefa.org.au/resources/icdgb/icdgb.htm#cont0 (n.d.)
- Notes From Nowhere, *We Are Everywhere: The Irresistible Rise of Global Anticapitalism* (London, 2003)
- Marcus Rediker and Peter Linebaugh, *The Many-Head Hydra: The Hidden History of the Revolutionary Atlantic* (London, 2002)
- T.V. Reed, *The Art of Protest: Culture and Activism from the Civil Rights Movement to the Streets of Seattle* (Minnesota, 2005)
- Aidan Ricketts, *The Activists Handbook* (London, 2013)
- Ruckus Society, *Action Strategy: A How-To Guide* (Oakland, n.d.)
- Seeds for Change, briefings and guides for actions, meetings, co-ops and campaigns: www.seedsforchange.org.uk/resources
- Gregory Sholette, *Dark Matter: Art and Politics in the Age of Enterprise Culture* (London, 2011)
- Stevphen Shukaitis and David Graeber (eds), *Constituent Imagination: Militant Investigation // Collective Theorization* (Edinburgh, 2009)
- Tactical Tech, *The Info-Activism How to Guide*, howto.informationactivism.org
- E.P. Thompson, *The Making of the English Working Class* (London, 1991)
- Trapese Collective, *Do It Yourself: A Handbook for Changing Our World* (London, 2007)
- Mark Traugott, *The Insurgent Barricade* (Berkeley, 2010)
- Turbulence Collective (ed.), *What Would It Mean to Win?* (Oakland, 2010)
- Raoul Vaneigem, *The Revolution of Everyday Life* (Oakland, 2012)
- Colin Ward, *Anarchism: A Very Short Introduction* (Oxford, 2004)
- Andrew X, 'Give Up Activism', *Do or Die: Voices from the Ecological Resistance 9* (Brighton, 2001), pp.160–6
- Howard Zinn, *A People's History of the United States: 1492 to the Present* (New York, 2005)

Unless otherwise stated, images are © V&A • Inside front cover Courtesy of Anıl Aykan • **Inside back cover** © Andrew Testa/Panos • **2** AFP/Getty Images • **6 (bottom)** Courtesy of María Teresa Nannini • **8 (top)** By permission of People's History Museum • **10–11** PA Images • **12** Alexander Turnbull Library, Wellington, New Zealand • **14** Edward Hasbrouck (Hasbrouck.org) • **17** Photograph by John M. Miller • **19 (bottom)** Photograph by Magdalena Jitrik, courtesy of *TPS* archive **(top)** © David Hoffman • **21** Free Content, free use information and non-commercial (Copyleft). Image downloadable from: n-1.cc/file/view/1869125/estamos-vivos-parece • **22** Imaginechina • **23** Courtesy of Pavel Arsenev • **28 (left)** © Bibliothèque nationale de France, 2013 • **33 (bottom)** REUTERS/Murad Sezer • **36 (bottom)** Courtesy of Fabian Frenzel **(top)** Free Content, free use information and non-commercial (Copyleft). Image downloadable from: n-1.cc/file/view/1869118/mapa-de-las-comisiones-de-la-acampada • **38** Immo Klink • **40** 123 Occupy • **41** Photograph by Ian Campbell • **43** Isy Morgenmuffel • **44–45** Photograph by Seamus Travers • **46 (bottom left, bottom right)** #occupygezi Architecture **(top right)** Tilly Gifford • **48 (bottom)** ASSOCIATED PRESS **(top)** PA Images • **50 (1)** Houghton Library, Harvard University **(2)** Modern Records Centre, University of Warwick **(3)** GD467/2/1/7, Records of Outright Scotland, National Records of Scotland **(5)** anticopyright **(6)** Photograph by Jeff Parks • **51** Occupy Sandy • **53** Immo Klink • **56** Vittorio Giannitelli • **57 (bottom right)** J. Boehnert, EcoLabs **(top left)** Oriana Eliçabe **(top right)** Immo Klink • **58 (bottom)** Indymedia **(top)** PA Images • **59** PA Images • **60–61** PA Images • **62 (top/bottom)** Oriana Eliçabe/Enmedio • **63** Courtesy of the Orange Alternative Foundation, www.orangealternativemuseum.pl • **64 (bottom)** © Andrew Testa/Panos **(top)** PA Images • **65** Lock-on Devices how-to based on drawings by B Dahl and North East Forests Alliance • **66** B Dahl • **67 (bottom left)** REX/Associated Newspapers **(bottom right)** Immo Klink **(top left)** Jarrah Keenan **(top right)** Photograph by Günter Zint • **70** Jonathan Slaff • **71** Riccardo Colella • **72** Friedel Fisher • **73 (bottom)** Oriana Eliçabe/enmedio.info **(top)** Photograph by Greg Sholette • **74** © Gareth Lind, 2001 • **76** Nick Cobbing/Alamy • **81, 82** Courtesy of Unpopular Books • **88** Courtesy of Blanca Garcia, the image on the banner is based on a poster designed by Matt Height • **89** ARTPLAY • **90** Alex Garland • **91** Ed Hall • **92 (top right)** Unite the Union • **94 (bottom)** By Andy Dao & Ivan Cash – 2011 • **95** Linen Hall Library, Belfast • **96** Korean Democracy Foundation Archives • **97 (top)** © Noel Douglas **(bottom)** Institute for Applied Autonomy • **98** London Greenpeace Group • **99 (top)** Photograph by Cat Mazza • **99** Knitting Pattern how-to based on designs by Cat Mazza • **101** Silvio Zuccheri • **102** Bruno Quaretti • **103** CAPATACO • **104–105** Courtesy of *Grupo Arte Callajero* • **109** Clandestine Syrian Stencil how-to based on designs by Malu Halasa, Zaher Omareen and Ibrahim Fakhri • **110–111** Donna Binder • **113** © Jérémie Dubé-Lavigne, Montréal, Québec, Canada; jeremiedubelavigne.4ormat.com • **117** Photograph by Mohammed Maree • **117** Bucket Pamphlet Bomb how-to based on drawings by Ken Keable • **118** PA Images • **120 (top)** Photograph by George Lange • **121 (bottom)** Sharon Hayashi **(top)** Masasit Mati • **123** Photograph by Martin Melaugh • **124 (bottom left)** Courtesy of the artist **(bottom middle)** Sue Ann Harkey **(bottom right)** Courtesy of the artist **(top)** Photograph by Paul Herrmann • **125 (top)** Photograph by Teresa Sordo Vilchis • **126 (left)** Sukey **(right)** Molleindustria • **127 (top)** Center for Tactical Magic **(bottom)** Photograph by Siobhan Mckeown • **128** Bike Bloc how-to based on designs by Chris Pepler, taken from initial concept by Bike Bloc Copenhagen and Sound Swarm Hamburg • **129** © Andrew Testa/Panos

oofs, tyres, photocopied pictures, carts,
ields, power chargers, automobiles, carr
otype machines, manure, caravans, dun
ooks, cameras, tarpaulins, banners, videc
aled envelopes, lamp posts, catalogues,
stols, paper, food, windows, empty water
aint, mattresses, Plexiglas shields, shopp
nces, acetates, handles, bodies, stamps,
olypropylene ropes, teacups, drainpipes,
bles, casements, forme, flatpacked pop-
dsteads, bodies, gaffer tape, doors, card
acksmith's anvils, paper, effigies, covers,
rses, media devices, traffic signs, letterp
shirts, stones, photocopiers, wood scrap
oxes, juggling clubs, walls, school aprons
dders, screen-print stencils, ring-bound
ousehold furnishings, bodies, helmets, sn
urnals, rulers, banisters, pneumatic drills
ooden and wire poles, public billboards,